EVOLUTION OF ELECTROCHEMISTRY

Key Figures and Their Contributions

Evgeny Katz

First Edition

NEWMAN SPRINGS PUBLISHING
320 Broad Street
Red Bank, NJ 07701

First originally published by Newman Springs Publishing 2024

ISBN 979-8-89308-937-0 (Paperback)
ISBN 979-8-89308-939-4 (Hardcover)
ISBN 979-8-89308-938-7 (Digital)

Printed in the United States of America

CONTENTS

PREFACE

While modern science and technology are studied at different levels in schools, colleges, and universities, the history of them—particularly personalized with scientists who contributed to their development—is frequently not known enough. The present book is aimed at filling a gap between modern knowledge and the history of its progress over time. The book structure is different from many other related textbooks and Internet websites; it gives more personalized overview of the knowledge progress rather than a scientific description of the achieved progress.

In the rush toward new research results and developments, the fundamental origins of the concepts and techniques employed in electrochemistry can be overlooked, perhaps most of all by researchers new to the field. What were the concepts that led to inventing and eventually producing the instrument standing on the benchtop? Why does a certain technique carry a particular name? Who was behind it, and why? Where did it all come from (P1)?

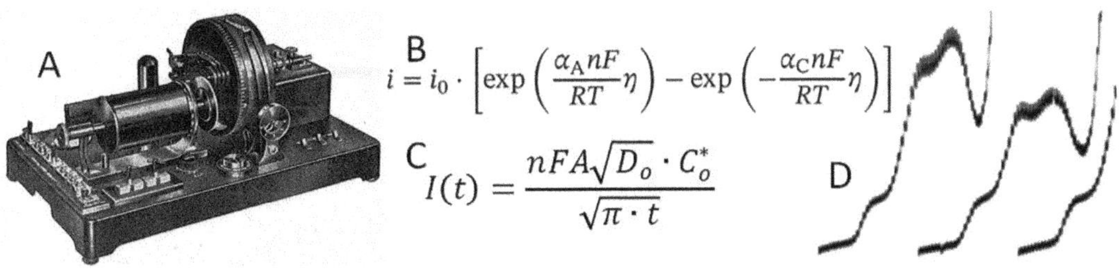

P1. (A) The first polarograph—the precursor of modern electrochemical analyzers. Can you believe that this mechanical machine is an electrochemical device? (B) and (C) Butler-Volmer equation and Cottrell equation, respectively. Do you know why these equations have these names? (D) Brdička waves—what do they mean, and why they have this name?

Understanding the roots of the techniques and concepts used in the lab can give researchers a wider intellectual base from which to reach for ever higher achievements. There are so many aspects and subfields in and near electrochemistry that it would require immense effort for any researcher to pursue all the relevant history individually.

Some personalities who greatly contributed to the study of electrochemistry are commonly known—for example, Jaroslav Heyrovský and Aleksandr Naumovich Frumkin—while many other names have been almost forgotten, such as Dionýz Ilkovič and Tatyana Aleksandrovna Kryukova (P2 and P3).

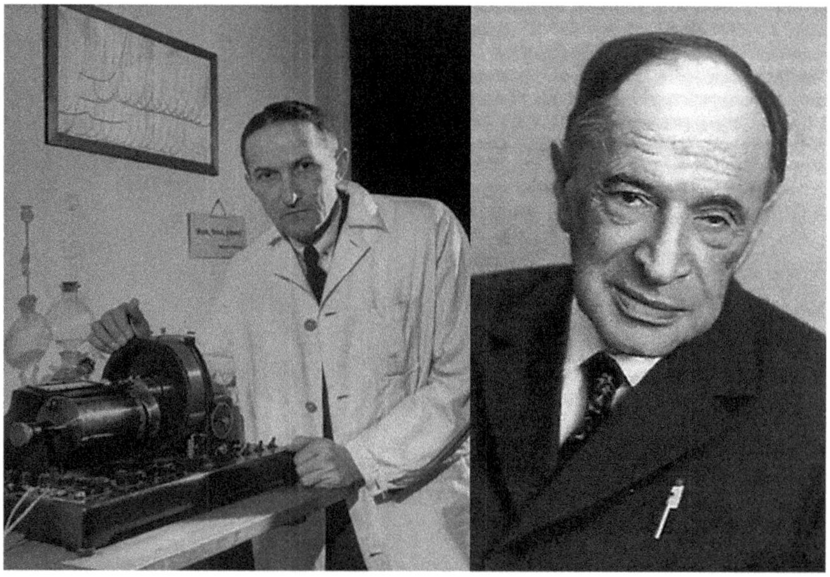

P2. Jaroslav Heyrovský with the first polarograph (*left*) and Aleksandr Naumovich Frumkin (*right*)

P3. Dionýz Ilkovič (*left*) and Tatyana Aleksandrovna Kryukova 1960s (*right*)

Information on the remembered and forgotten personalities can be relatively easily found on the Internet; however, in order to find this information, one has to know about whom the search should be done. Then the lack of the initial information does not allow us to find any data related to a specific person and his/her contribution to science and technology and particularly to electrochemistry.

The purpose of this book is to offer readers a short summary for a selection of the true breakthroughs and historic contributions that have shaped electrochemistry over the years. There may even be a few surprises along the way and, hopefully, ample inspiration for readers to look more deeply. The book does not pretend to be an encyclopedia of electrochemistry; otherwise, it would have several volumes and thousands of pages. The selection of the personalities included in the book is mostly based on the personal preferences of the author. Particularly, the scientists who contributed to polarography (the electroanalytical technique using a dropping mercury electrode) are highlighted more than scientists who worked on the corrosion problems. This selection originated from the personal scientific experience of the author, who started his scientific career from performing polarography and cyclic voltammetry experiments (P4), but never worked on corrosion.

P4. Evgeny Katz, a PhD student (*left*); Yury Kozlov (*right*); and Dr. Boris Kiselev (*sitting*) performing cyclic voltammetry experiments (Institute of Photosynthesis, USSR Academy of Sciences, Pushchino, Moscow Region, 1977)

The format of the book is chronological according to the years when scientists were born. Obviously, this chronological order cannot be the same as the time or time period of the specific discovery or invention.

The book does not pretend to be read from the first page to the end. It is mostly a database on scientists, thus allowing the readers to find more information about the briefly described scientists, if more information is needed. Carefully, composed indexes placed at the end of the book allow finding specific persons or things (devices, processes, concepts, inventions, etc.). This book took a lot of time to be written, being a hobby for the author rather than his scientific assignment, which is in the areas of bioelectronics and bionanotechnology—areas not really related to this book.

I would like to conclude this preface by thanking my wife, Nina, for her support in every respect in the past fifty-three years, from the time when I was a student to the present time when I am close to retirement. Without her help, it would not have been possible to complete this work (P5).

P5. Evgeny and Nina Katz at a scientific conference

FIGURES, THEIR SELECTION, AND COPYRIGHT PERMISSIONS

All figures illustrating the text were carefully selected from numerous images available on the Internet. The selection was based on their perfect illustration of the text and on their copyright status to allow their reproduction in the book. The great majority of the selected figures are copyright-free (public domain), then allowing their reproduction without requesting copyright permissions. Particularly, some figures were copied from old books, journals, catalogs, patents, and newspapers for which copyrights already expired. Reproduction of a small fraction of the used figures required permissions, which were obtained from the copyright owners. A few schemes were prepared by the author. They were redrawn to allow better graphical quality and higher resolution compared to the low-quality schemes available on the Internet.

1

LUIGI GALVANI

(Latin: *Aloysius Galvanus*)

Born: September 9, 1737, Bologna, Papal States (presently Italy)

Died: December 4, 1798, Bologna, Cisalpine Republic (presently Italy)

Luigi Galvani (1.1) was an Italian physician, physicist, biologist, and philosopher who investigated the nature and effects of what he conceived to be electricity in animal tissue. Electrochemistry and particularly bioelectrochemistry and bioelectronics have been started from Galvani's experiments.

In numerous carefully performed experiments, Luigi Galvani—assisted by his wife, Lucia—discovered that the muscles of dead frogs' legs twitched when they are touched with electrified metal instruments (1.2A). The experimental conditions were different but always related to electricity (1.3A). For example, he obtained muscular contraction in a frog by touching its nerves with a pair of scissors during an electrical storm (1.3B). In another experiment, the electrical effect was obtained while an electrical machine was activated (1.4A). Galvani also demonstrated the twitching without the aid of an electrostatic machine while touching a dead frog with two different metals simultaneously, thus discovering what we presently know as formation of a Galvanic cell for producing electricity due to contacting metals with different redox potentials.

Galvani delayed the announcement of his findings until 1791, when he published his essay "*De Viribus Electricitatis in Motu Musculari Commentarius*" ("Commentary on the Effect of Electricity on Muscular Motion") (1.2B). He theoreticized that the electrical effect on the biological tissue relates to the natural electricity existing in the tissue preserved even in the dead frogs. He termed it "animal electricity," which is naturally produced through mechanisms different from another *natural* form of electricity generated by lightning or by electrostatic friction machines, which were the only laboratory generators of electricity at his time. He considered the brain as the most important organ for the secretion of the electric fluid and the nerves operating as conductors delivering the electricity to muscles. Galvani tried to explain the muscle operation similarly to charging the Leyden jar but producing the charges exclusively by biological processes. Importantly, Galvani's theoretical approach was based on the concept that the electricity is generated by biological processes.

A part of the scientific community accepted Galvani's theory of "biological electricity," but Alessandro Volta (Italian physicist and chemist, 1745–1827), the outstanding professor of physics at the University of Pavia, suggested another explanation based on physical rather than biological origin of electrical observations reported by Galvani. In his explanation, the frog's legs operated only as indicators of electricity produced in some physical processes similar to electroscope leaves.

These two confronting concepts of biological or physical origin of electricity resulted in a long-term scientific discussion between Galvani and Volta followers. As a result of this scientific discussion, in 1794, Galvani offered a defense of his position in an anonymous book titled *Dell'uso e dell'attività dell'arco conduttore nella contrazione dei muscoli* (*On the Use and Activity of the Conductive Arch in the Contraction of Muscles*), the supplement of which described muscular contraction, exposing frog's muscle to another biological object only (with no use of any metal). He caused a muscle to contract by touching the exposed muscle of one frog with a nerve of another and thus established for the first time that bioelectric forces exist within living tissue. As it frequently happens in science, both theories were partially right and wrong as can be seen from the present scientific knowledge.

Galvani—as well as his opponent Volta—performed numerous public experiments to promote the electric research. It was quite common that time to perform scientific experiments in public (1.4B), particularly in aristocratic salons. The scientific contribution of Galvani, as well as Volta, has been acknowledged by introducing many electrical terms with his name: *galvanic element*, *galvanic process*, *galvanization*, etc.

1.1. Luigi Galvani (oil-painting, unknown artist)

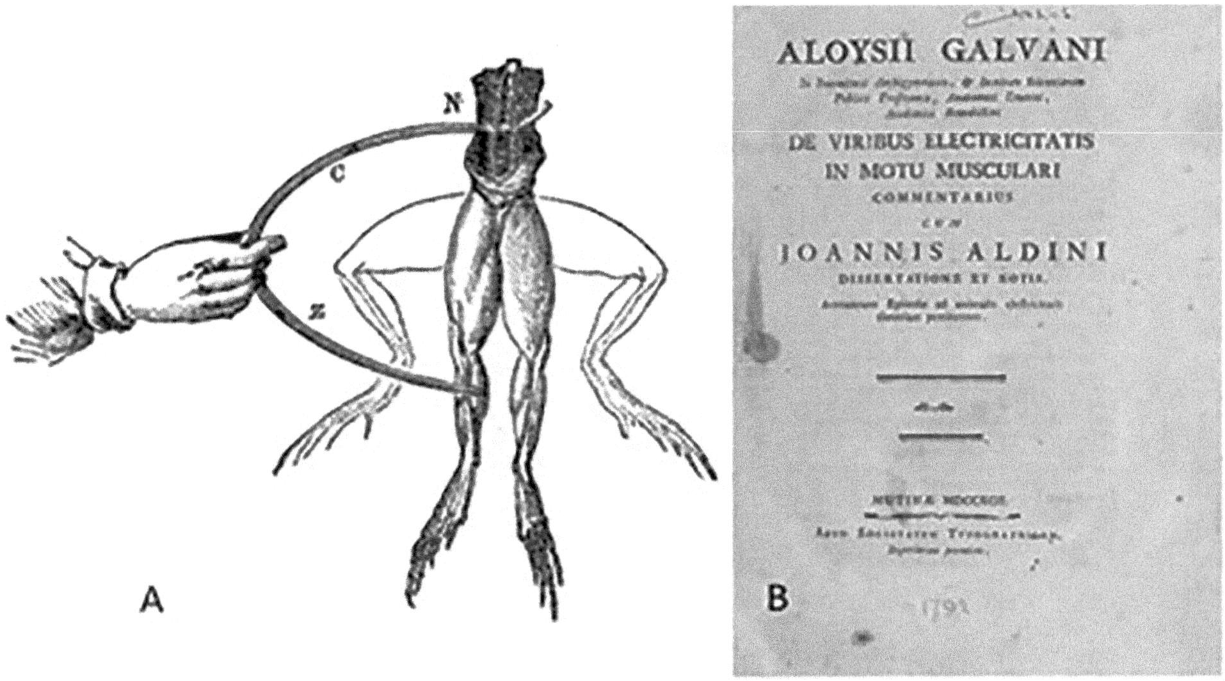

1.2. (A) Galvani's experiment with frog's legs: electrodes touch a frog and the legs twitch into the upward position. (B) The title page of Galvani's book: *De Viribus Electricitatis in Motu Musculari Commentarius*

A B

1.3. (A) Galvani's experiments on frog's legs—various experimental conditions outlined in Galvani's book. (B) Late 1780s diagram of Galvani's experiment on frog's legs. In this experiment, the electricity was delivered to a dead frog from an electric storm. The drawing originates from Galvani's publication, "*De viribus electricitatis in motu musculari commentaries.*"

A B

1.4. (A) Luigi Galvani experiment on frog's legs (originated from Galvani's publication, "*De viribus electricitatis in motu musculari commentaries*"). This engraving shows the electrical apparatus of Luigi Galvani used to produce an electric cell with the fluids extracted from the body of a frog. (B) Galvani demonstrating in public his famous experiment with frog's legs. An electrostatic machine is shown on the table (artwork from *Electricite* [1911] by Max de Nansouty, part of the *Les merveilles de la science* series of 1867–1891 by Louis Figuier).

2

ALESSANDRO GIUSEPPE ANTONIO ANASTASIO VOLTA

Born: February 18, 1745, Como, Lombardy, Italy
Died: March 5, 1827, Camnago Volta, Italy

Alessandro Giuseppe Antonio Anastasio Volta (2.1) was an Italian physicist, chemist, and pioneer of electricity who is credited as the inventor of the first electric battery. He invented his battery in 1799 and reported the results of his experiments in 1800 in a two-part letter to the president of the Royal Society.

The battery made by Volta (named Voltaic pile) is credited as one of the first electrochemical cells. It consists of two electrodes: one made of zinc (Zn) and another of copper (Cu). (Volta experimented with many other metals, but Zn and Cu were the most frequently used.) In order to organize a solid-state device, Volta used cloth dampened with brine (a salted water solution) or with diluted sulfuric acid between the metallic electrodes. In his explanations, Volta considered the contact between two different metals as the source of electricity.

In reality, according to the present knowledge of electrochemistry, the mechanism of the battery operation was different. When the Zn and Cu electrodes were electrically connected, the Zn electrode was oxidized to form Zn^{2+} cations in the solution, donating electrons to the electric circuit. Meanwhile, H^+ ions were reduced at the Cu electrode, accepting electrons and producing H_2 gas. Copper metal did not react but provided a conducting platform for the H^+ reduction. Overall, the system generated voltage and current between the Zn and Cu electrodes. According to modern terminology, the Zn electrode operated as an anode while the Cu electrode was a cathode. The voltage generated between them was determined by the difference of the redox potentials of both electrochemical reactions.

Schematically, the voltaic pile operation can be presented by the following reactions, using modern notations:

$Zn \rightarrow Zn^{2+} + 2\,e^-$ (anodic reaction)
$2\,H^+ + 2\,e^- \rightarrow H_2$ (cathodic reaction)
$Zn + 2\,H^+ \rightarrow Zn^{2+} + H_2$ (global reaction)
(anode: oxidation) $Zn \mid Zn^{2+} \parallel 2H^+ \mid H_2 \mid Cu$ (cathode: reduction)

It should be noted that Volta did not use this form of presenting the chemical processes, and his explanation of the electricity formation was different. He believed that the electricity is produced due to the contact between two metals while he did not take into account the electrolyte reaction.

In order to increase the voltage produced by the system, Volta assembled his battery from many pairs of the Zn-Cu electrodes connected in series (2.2). The voltage produced by the battery was proportional to the number of the electrode pairs. With this invention, Volta proved that electricity can be generated chemically, arguing with Luigi Galvani (Italian physician, physicist, biologist, and philosopher, 1737–1798) who believed in the biological origin of electricity.

It is interesting to note that Volta has demonstrated his battery and some electric experiments for Napoleon Bonaparte, who was deeply impressed by the demonstrations and then supported Volta's research (2.3).

The voltaic pile became one of the first reliable sources of electricity, and that new source of study enabled many discoveries in the field, from understanding of the electrolysis of water to the research of electric arcs. Volta's experiments, particularly his battery—were highly important for electric studies and led to the development of the field of electrochemistry. Volta made many other important inventions and discoveries (e.g., he discovered methane), but his contribution to electrochemical studies was the most important. For recognizing his tremendous contribution to electrical studies, including electrochemistry, a unit for measuring electric potentials was named after him as the Volt (V).

2.1. Alessandro Giuseppe Antonio Anastasio Volta (portrait by an unknown artist with a voltaic pile shown in the right foreground)

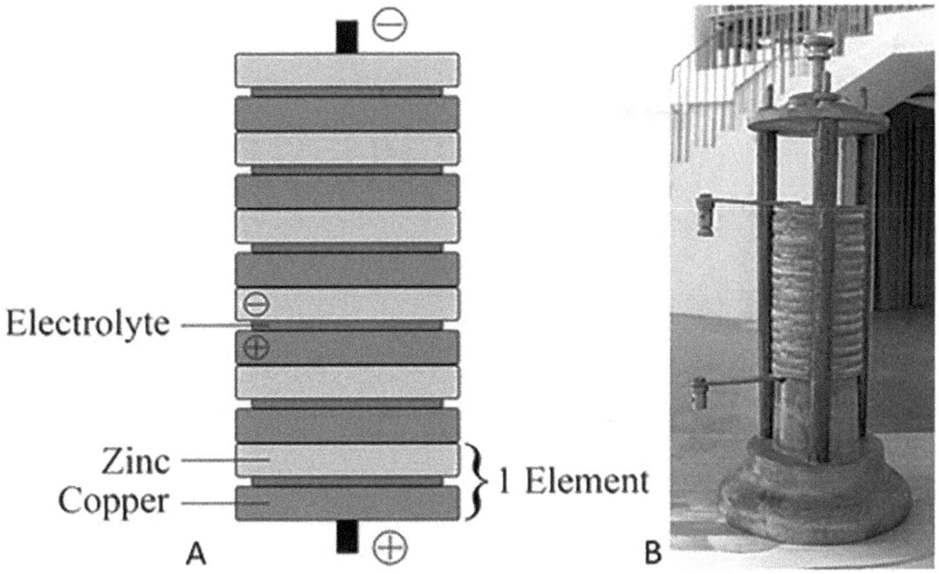

2.2. (A) Schematics of the Voltaic pile composed of many Zn/Cu elements; (B) A voltaic pile device (reconstruction)

2.3. Volta demonstrating his battery and performing electric experiments for Napoleon Bonaparte (sitting) (1801). (Painting by an unknown artist)

3

WILLIAM NICHOLSON

Born: December 13, 1753, London, England
Died: May 21, 1815, Bloomsbury, London, England

William Nicholson (3.1) was an English chemist, journalist, publisher, inventor, patent agent, and civil engineer. He wrote several important books on very different topics, ranging from physics and chemistry to architecture and marine navigation. His important contribution to science was establishing in 1797 the first periodic scientific journal, *Journal of Natural Philosophy, Chemistry, and the Arts*, which was generally known as Nicholson's Journal. The success of this periodical inspired the creation of several scientific journals in England. Nicholson's most important experimental achievement was the discovery of water electrolysis.

In 1800, after he heard of the invention of the electric battery by the Italian physicist Alessandro Volta, Nicholson constructed a similar battery and performed a number of experiments with it. While doing various experiments with the voltaic pile (the Zn-Cu battery invented by Volta), Nicholson, with his friend Anthony Carlisle, found that when the battery poles are immersed in water, they produce bubbles of gas identified as oxygen (O_2) and hydrogen (H_2) (3.2). This discovery was conceptually important because it demonstrated for the first time the interaction of electricity and chemical materials, resulting in their chemical transformations (3.3). In other words, they produced a chemical process stimulated by electric energy. Eventually, the whole of electrochemistry started from this experiment.

3.1. Portrait of William Nicholson, ca. 1812, engraved by Thomas Blood after a portrait painted by Samuel Drummond (1765–1844)

3.2. The drawing shows Anthony Carlisle (left) and William Nicholson doing an experiment with electricity. Three sources of electricity are shown: a voltaic pile, a frictional electrical generator, and two Leyden jars. The picture originates from the book *Les Merveilles de la Science* by Louis Figuier. The original caption in the French book claims that Carlisle and Nicholson are doing an experiment with water electrolysis; however, upon careful examination of the image, one can see that they are measuring voltage of a volta pile with a gold-leaf electrometer.

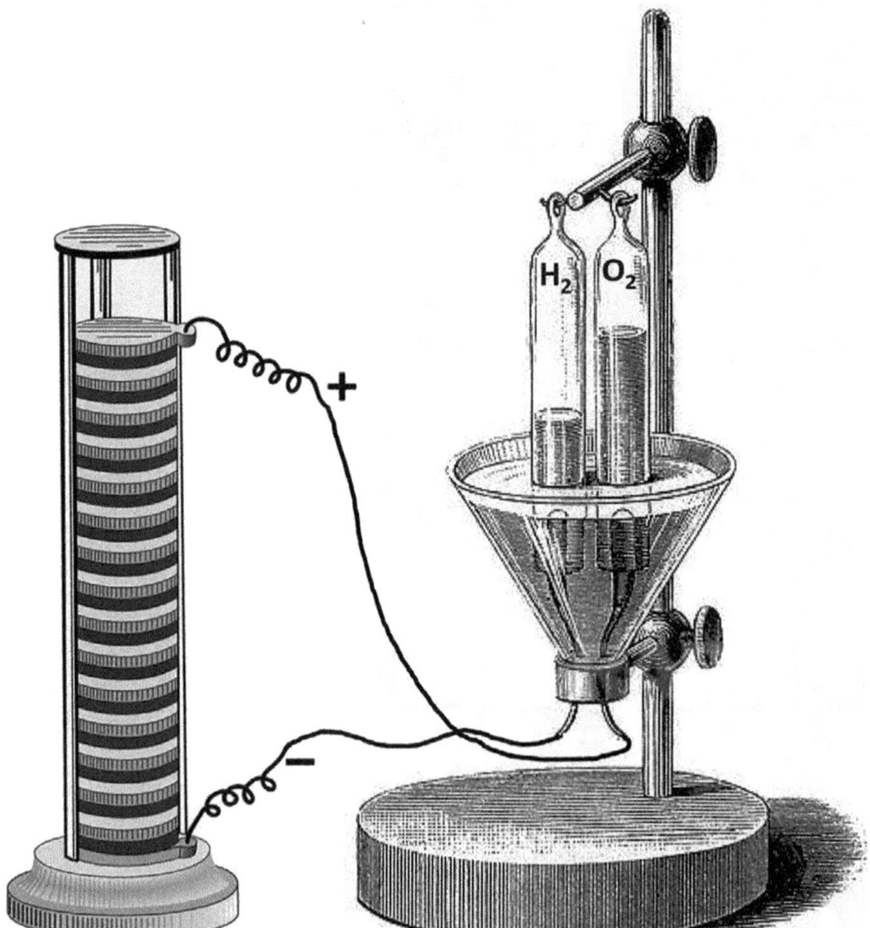

3.3. The scheme shows the setup used by Nicholson and Carlisle for their experiments with water electrolysis. The device included a voltaic pile (battery) connected to the electrodes immersed in water. The produced gases were collected, and their volumes were measured in the proportion of H_2:O_2 = 2:1, thus confirming the water composition H_2O. Notably, O_2 and H_2 were produced at the positive and negative poles of the battery, respectively.

4

HUMPHRY DAVY

Born: December, 17, 1778, Penzance, Cornwall, Great Britain
Died: May 29, 1829, Geneva, Switzerland

Humphry Davy was a Cornish chemist and inventor (4.1). The Cornish people belong to a Celtic ethnic group, which is a minority in the United Kingdom, tracing its roots to the ancient Britons populated Great Britain before the Roman conquest. Davy is best known for isolating, by using electrolysis, many elements for the first time—potassium (K) and sodium (Na) in 1807 and calcium (Ca), strontium (Sr), barium (Ba), and magnesium (Mg) in 1808. Actually, Davy was one of the originators of electrochemistry as a subarea of chemistry. With his experimental work, Davy proved the practical use of electrochemistry for obtaining many different chemical materials and compounds (4.2).

For his electrochemical (electrolysis) experiments, Davy designed very powerful electric batteries composed of many galvanic elements connected in a circuit (4.3). Notably, the first of Davy's attempts to produce alkali (Na, K) and alkaline earth (Mg, Ca, Sr, Ba) metals by electrolysis of their salt solutions in water were unsuccessful. Only hydrogen (H_2) was produced at the negatively polarized cathode. It is not surprising with the present knowledge of redox potentials of these metals, which are more negative than the hydrogen potential. However, this was not known at Davy's time. Then experimenting with molten salts (excluding water), Davy succeeded in producing active metals, which cannot be produced electrochemically from aqueous solutions.

In 1815, Davy suggested a theory explaining the composition and properties of acids and bases. According to his theory, acids were substances that contained hydrogen ions (H^+) replaceable partially or totally by metals placed above hydrogen in the reactivity series (historic analog of presently used redox potentials). When acids reacted with metals, they formed salts and hydrogen gas. Bases were substances that reacted with acids to form salts and water. These definitions worked well for most of the nineteenth century. They are partially correct now.

Among many other studies and discoveries, in 1799, Davy experimented with nitrous oxide (N_2O) and discovered its anesthetic properties in relieving pain during surgery. Davy coined its name as "laughing gas" due to its euphoric effect upon inhalation. It is interesting to note that the laughing gas has found its recreational use, similar to mild narcotics. One of Davy's important practical inventions was a miner's safety lamp. Overall, Davy's research on electrolysis, particularly of molten salts, is his most important contribution to electrochemistry.

4.1. Portrait of Sir Humphry Davy (painted by Thomas Phillips, 1770–1845, a leading English portrait painter)

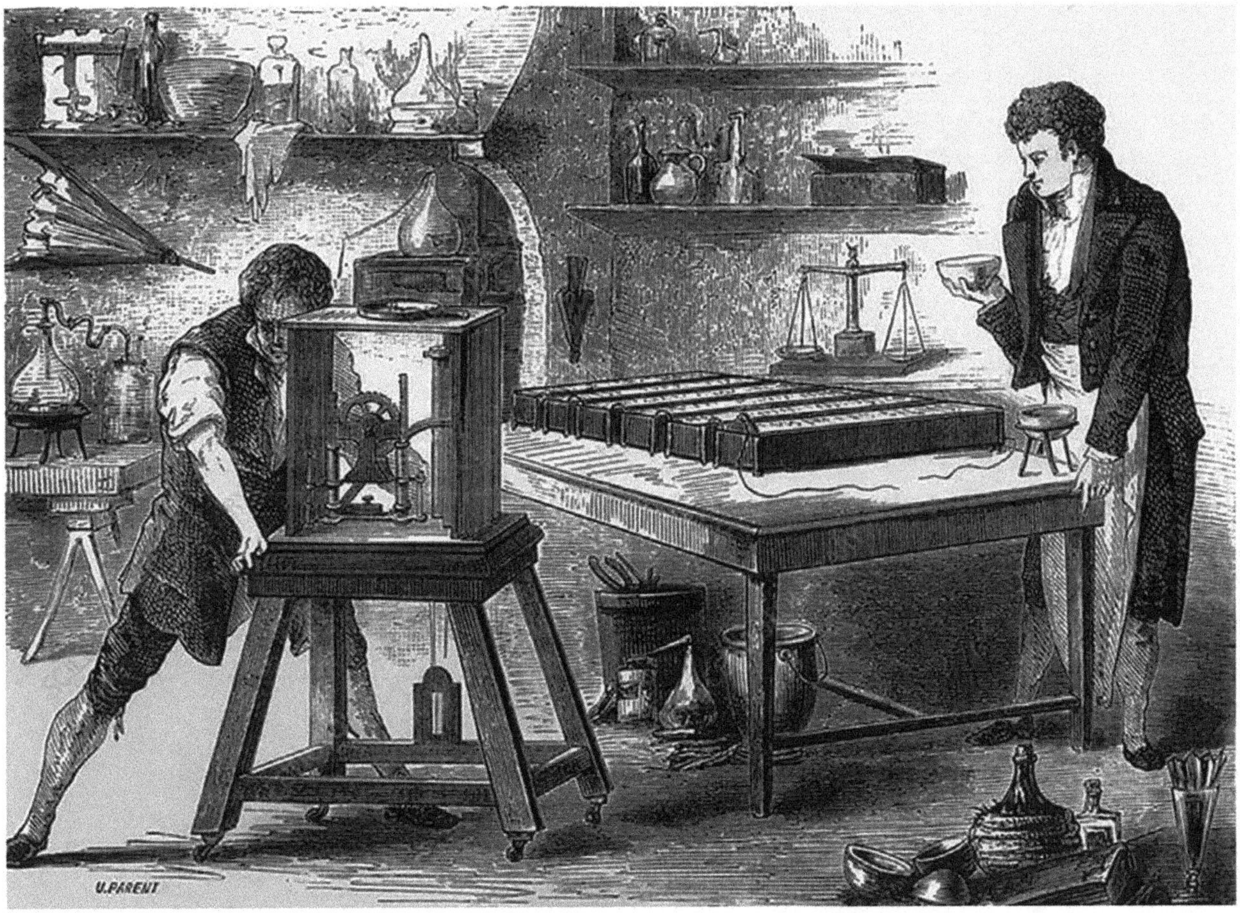

4.2. Davy using a voltaic battery to experiment with the decomposition of alkalis. The picture shows Davy (shown at the right) with an assistant experimenting at the Royal Institute in 1807. The voltaic batteries are seen on the table (engraving by an unknown artist).

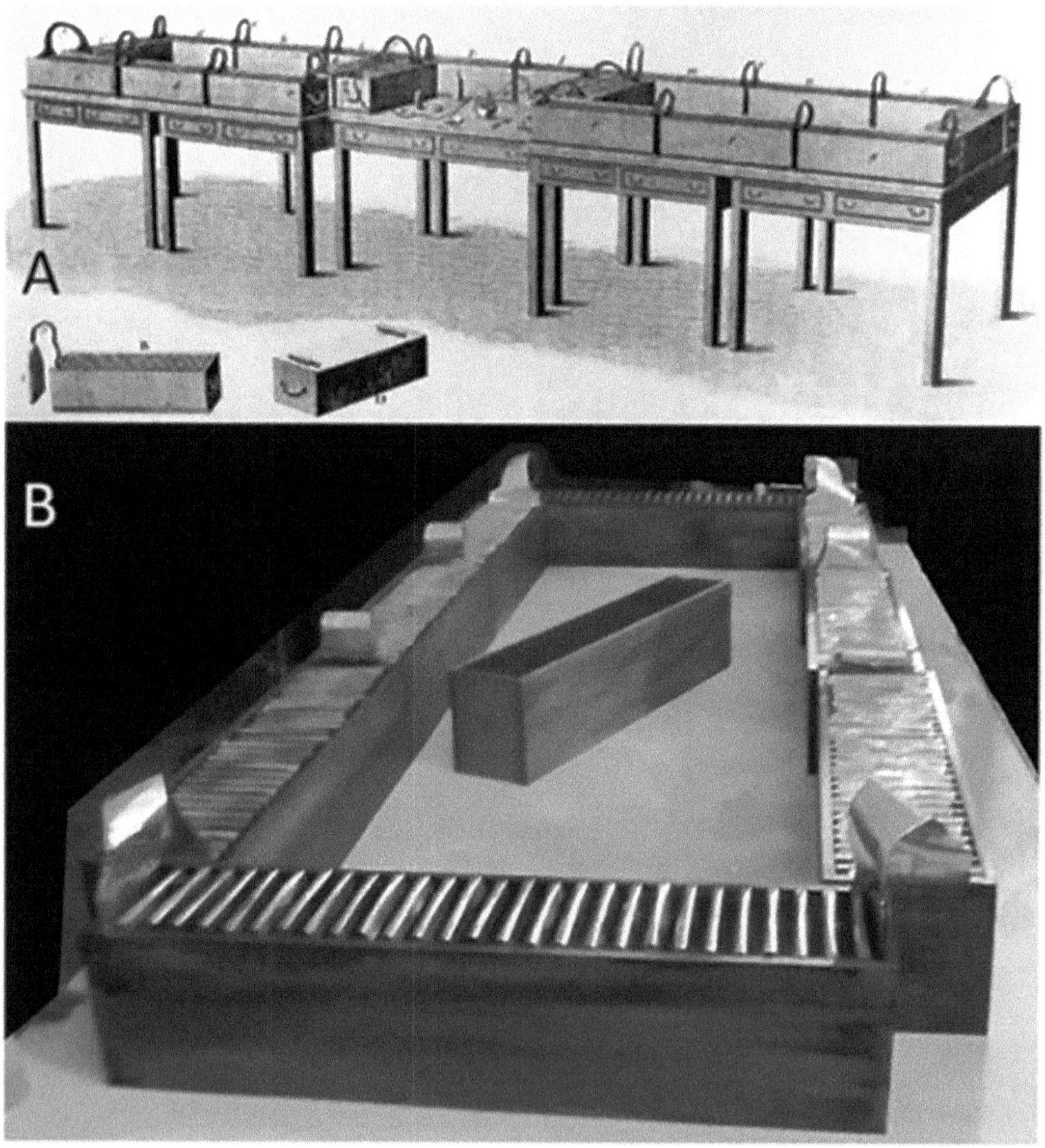

4.3. Davy's electric battery composed of many individual sections connected in circuit: (A) original engraving and (B) replica provided up to 200 V. A similar battery was used by Davy for electrolytic production of potassium.

5

JÖNS JACOB BERZELIUS

Born: August, 20, 1779, near Linköping, Sweden

Died: August, 7, 1848, Stockholm, Sweden

Jacob Berzelius (formal name: Baron Jöns Jacob Berzelius) (5.1) was a Swedish chemist. He is considered—along with Robert Boyle, John Dalton, and Antoine Lavoisier—to be one of the founders of modern chemistry. His most important contribution to chemistry was in the areas of determination of atomic weights, stoichiometry, chemical bonding, and electrochemistry. Berzelius developed novel efficient methods of chemical analysis, investigated isomerism, allotropy, and catalysis.

By 1818, Berzelius had obtained, with a high degree of accuracy, the atomic weights of forty elements. He discovered new elements, cerium (^{58}Ce) and selenium (^{34}Se), isolated in pure form silicon (^{14}Si) and thorium (^{90}Th) and formulated the difference between inorganic and organic compounds. Berzelius introduced into chemistry the chemical formula notation showing the composition of compounds. He used abbreviated Latin names of the elements (taking one or two letters from the Latin names) and numbers of atoms of each element present (e.g., Na_2SO_4 for sodium sulfate; note the Latin names *natrium* for sodium, *sulphurium* for sulfur, *oxygenium* for oxygen). Note that Berzelius's original chemical nomenclature was slightly different from the modern version, but it has the same meaning.

Berzelius conducted pioneering experiments in electrochemistry, resulting in establishing the law of constant proportions, which states that the elements in inorganic substances are bound to each other in definite proportions by weight. Using the electrical battery designed by Volta (voltaic pile), Berzelius studied decomposition (*electrolysis* in the modern definition) of various inorganic compounds. He observed that many compounds could be decomposed by an electric current in an acidic component at the positive pole and a basic component at the negative pole.

Based on the experimental results, Berzelius formulated electrochemical dualism theory, suggesting that all compounds are actually salts of two opposite primary materials—one of acidic and another of basic nature. For example, potassium sulfate (K_2SO_4) was considered by him as a salt produced by the reaction of a basic K_2O and an acidic SO_3. According to Berzelius, all chemicals—whether natural or artificial, mineral or organic—could be distinguished and specified qualitatively by identifying their electrically opposing constituents (acids and bases in his formulation). While from the modern point of view, this theory is incorrect (better say naive). At his time, it was an important step to understand the composition of compounds. Berzelius also coined the term *protein* while applying his dualistic organic chemistry to the constituents of biological substances. He derived it from the Greek word *proteios* (*primitive*), emphasizing the primary importance of these biochemicals for life processes.

Although many of Berzelius's ideas were eventually modified or proved wrong, he made unparalleled contribution to chemistry in general and to electrochemistry in part.

5.1. Jöns Jacob Berzelius (lithography portrait from 1836 made by P. H. van den Heuvell)

6

CHRISTIAN JOHANN DIETRICH (LATER THEODOR) GROTTHUSS

Born: January, 20, 1785, Geddutz, Kurland (now Gedučiai, Lithuania)
Died: March, 26, 1822, Geddutz, Kurland (now Gedučiai, Lithuania)

Theodor von Grotthuss (6.1) was a Lithuanian-born chemist of German ethnicity, known for establishing the first theory of electrolysis in 1805 and formulating the first law of photochemistry in 1817. After Volta invented his electric battery (voltaic pile), many scientists used it in various experiments, including electrochemical studies, particularly those related to electrolysis of aqueous solutions. Grotthuss actively contributed to this area both in terms of electrolysis experiments and particularly their interpretation. He proposed the mechanism of electric conductivity in water based on hopping of hydrogen ions (H^+) from one water molecule to another one, finally generating a chain of the charge transfer, which later named the Grotthuss mechanism (6.2). This first study, titled "On Decomposition of Water and Substance Dissolved in It under the Action of Galvanic Electricity," was published in French in Rome at the end of 1805 and made him famous. The results of Grotthuss's experiments and theoretical considerations were published in different European journals in various languages. In addition to the water electrolysis, Grotthuss performed very first studies in organic electrochemistry.

Grotthuss made important contributions to the analysis of photochemical processes. He established that only absorbed light would cause chemical transformations. He also noticed the proportionality of the photochemical effect to the time of irradiation. While presently these conclusions can be considered as trivial, they were highly important and not obvious at his time. Important results of photochemical studies were also published by Grotthuss in 1815 in one of his earlier papers, "On an Absorbent of Light and Some General Considerations Concerning the Phosphorescence and Colors." The Grotthuss observation that the photochemical reactions are driven only by the absorbed light was independently rediscovered by John William Draper (English scientist, 1811–1882) in 1842 and is known as the Grotthuss-Draper law. In addition to his major research in the areas of electrochemistry (particularly electrolysis) and photochemistry, Grotthuss studied combustion of gas mixtures, which was carried out in 1809–1812.

6.1. Theodor von Grotthuss (portrait by an unknown artist)

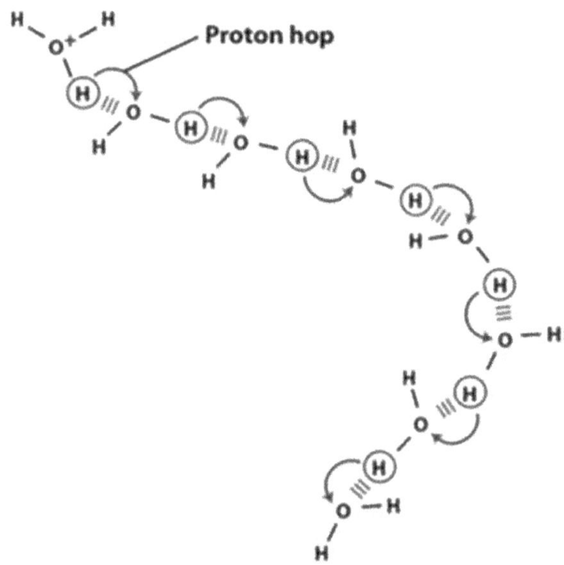

6.2. The Grotthuss mechanism of the charge transport in water based on hopping of H^+ species from one water molecule to the next one, then producing a chain of the charge transfer. The dash lines correspond to the hydrogen bonding between the water molecules.

7

ANTOINE CÉSAR BECQUEREL

Born: March 7, 1788, Châtillon-sur-Loing (now Châtillon-Coligny), France
Died: January 18, 1878, Paris, France

Antoine César Becquerel (7.1) was a French scientist and a pioneer in the study of many electric and electrochemical phenomena.

Becquerel performed experiments on the production of different materials by electrolysis, particularly producing sulfur and various metal sulfides by electrochemical means. He also produced metals from their ores, applying electrolysis. Among other achievements, in 1825, Becquerel invented a differential galvanometer for the accurate measurements of electric currents. In 1829, he developed the constant current cell (a battery) but received recognition for this cell only in France; elsewhere, the almost identical cell is called Daniell cell, acknowledging the invention by John Frederic Daniell (English chemist and physicist, 1790–1845).

In 1839, working with his son Alexandre-Edmond Becquerel, he discovered the photovoltaic effect on electrodes immersed in an electrolyte solution. While experimenting with metal electrodes and electrolytes, Becquerel discovered that conductance would increase with illumination.

It is interesting to note that Becquerel observed electrocapillary effect, which was not recognized by him as a highly important one. However, much later, in the 1920s, this effect resulted in the discovery of polarography, which was the very first electroanalytical method for which Jaroslav Heyrovský (Czech chemist and inventor, 1890–1967) received a Nobel Prize in Chemistry in 1959.

Overall, Antoine César Becquerel was the beginner of many important concepts and methods (particularly, piezoelectricity, electrolytic refining of metals, electroanalytical chemistry, etc.). However, he was not always able to bring them to the final formulation; thus, many credits for discoveries originally made by him have been given to other scientists.

7.1. Antoine César Becquerel (photo portrait, before 1878, by Petit Pierre, 1832–1909, photographer)

8

JOHN FREDERIC DANIELL

Born: March 12, 1790, London, United Kingdom
Died: March 13, 1845, London, United Kingdom

John Frederic Daniell (8.1) was an English chemist and physicist. His name is best known for the invention of Daniell cell, an electric battery much better than used at that time voltaic cells.

Daniell was a professor of physics at the University of Edinburgh in Scotland. He also worked as a chemist at the university and performed research in meteorology. In the 1830s, Daniell—under the influence of his friend Michael Faraday (English scientist, 1791–1867) (8.2A)—became interested in electric studies and concentrated on the development of an electric battery, trying to improve the voltaic pile (a battery named after Alessandro Giuseppe Antonio Anastasio Volta, an Italian scientist who invented this battery) used at that time. A major problem of the voltaic pile was its unstable performance. It was not capable of producing electric current/voltage in a sustainable way for a long period of time. This reduced its applicability for most experiments with electricity.

One of the problems limiting the stable operation of the voltaic pile was electrochemical corrosion of the negative Zn electrodes (operating as the anodes), particularly increased due to presence of different metal impurities in these electrodes. Actually, the presence of other metals in the Zn electrode produced internal shortcut electric circuits, reducing the voltage produced by the battery. This problem was solved, at least partially, by William Sturgeon (English physicist, 1783–1850), who amalgamated zinc electrodes, resulting in their longer and more stable operation.

Another problem was in the formation of a thin film of hydrogen bubbles at the positive Cu electrodes (operating as the cathodes), which increased the interfacial resistance and inhibited the voltage/current formation. Daniell began experiments in 1835 in an attempt to improve the voltaic battery, increasing its durability. While the original Volta's cell included a diluted sulfuric acid solution as an electrolyte, Daniell applied zinc sulfate ($ZnSO_4$) and copper sulfate ($CuSO_4$) solutions for the Zn and Cu electrodes respectively, separated by a porous ceramic spacer (8.2B).

The electrochemical process resulted in the anodic dissolution of the Zn electrode and cathodic deposition of metallic Cu on the Cu electrode. In order to keep the Cu^{2+} cations at the constant concentration in the electrolyte solution (important for preserving the constant voltage output), solid $CuSO_4$ salt crystals were added to the battery. They were used to maintain the saturated concentration of $CuSO_4$. Daniell battery operated with well reproducible voltage and current, much better than the original Voltaic pile, thus providing reliable source of electric power for many practical applications—particularly in early telegraph systems employing the Morse code (named after Samuel Finley Breese Morse, American inventor 1791–1872) (8.3).

Later, several improvements have been done for improving the battery operation, but all of them were based on Daniell's concept of the $Zn|Zn^{2+}||Cu|Cu^{2+}$ electric cell.

In 1839, Daniell produced a strong electric arc with a seventy-cell battery. Daniell also invented dew-point hygrometer (an instrument for measuring air humidity) and a pyrometer (a remote-sensing thermometer used to measure the temperature of distant objects). However, his name is mostly associated with the battery invented by him.

8.1. John Frederic Daniell (oil on canvas, unknown artist)

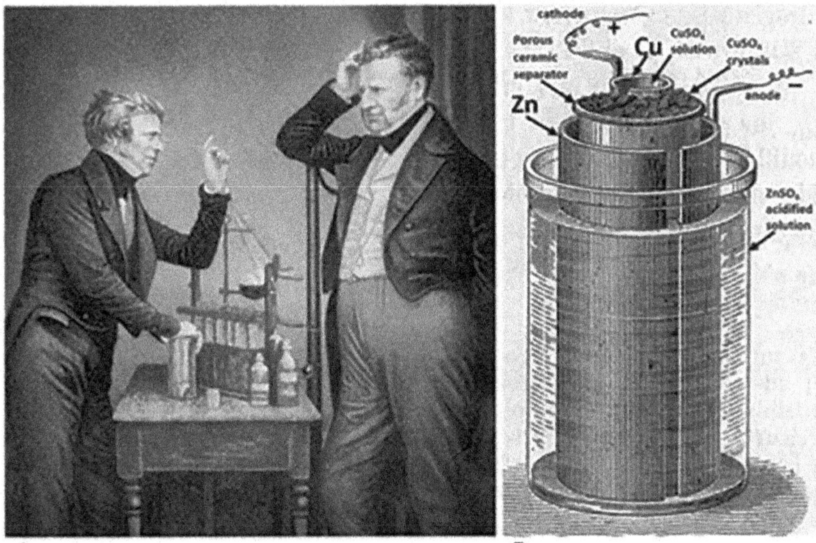

A B

8.2. (A) John Frederic Daniell (right) is shown with his friend Michael Faraday (left) doing electric experiments (engraving, 1848, by George Barclay, English printmaker, 1830–1872). (B) The original drawing of Daniell's battery (with added notations)

8.3. An illustration showing the inside of a rural telegraph station with a female operator. The depiction of the telegraphic equipment is accurate; on the operator's table are a telegraphic relay, a printing register, a cutoff switch, and a camelback telegraph key. On the floor is a battery box containing two Daniell cells, which supply electricity for the telegraph.

9

CHRISTIAN FRIEDRICH SCHÖNBEIN

Born: October 18, 1799, Metzingen, Duchy of Württemberg (presently Germany)
Died: August 29, 1868, Baden-Baden, Germany

Christian Friedrich Schönbein (9.1) was a German-Swiss chemist who is best known for experimental demonstration of a fuel cell concept, discovery of guncotton (nitrocellulose explosive), and study of ozone. The basic principle of later invented fuel cells was described in 1801 by Humphry Davy (British chemist and inventor, 1778–1829). The first experimental demonstration of the fuel cell operation was performed by Schönbein in 1838 and published in the January 1839 edition of the *Philosophical Magazine*.

He detected a voltage by dipping two platinum wires into a sulfuric acid electrolyte and applying flows of hydrogen and oxygen gases around them. It should be noted that Schönbein was only demonstrating the concept while much more work was done after him by Sir William Grove (British physicist, 1811–1896). Grove was a more practice-oriented man; he intended to use the fuel cell effect and designed a gas battery between 1842 and 1845. Therefore, the major credit for inventing fuel cells has been given to him while the original work of Schönbein was forgotten.

Notably, the *fuel cell* term was not used at that time, and it was coined later in 1889 by Ludwig Mond (British chemist and industrialist, 1839–1909) and Charles Langer (his assistant). Schönbein was a prolific writer, 364 papers appearing under his name in the Royal Society's catalogue, and he carried on a large correspondence with other men of science, such as Berzelius, Faraday, Liebig, and Wohler.

9.1. Christian Friedrich Schönbein (lithography by Rudolf Hoffmann, 1820–1882, after a photograph by Hanfstaengl, 1858)

10

GUSTAV THEODOR FECHNER

Born: April 19, 1801, Żarki Wielkie, Poland
Died: November 18, 1887, Leipzig, Germany

Gustav Theodor Fechner (10.1) was a German physicist, psychologist, and philosopher. His scientific contributions were mostly in the area of experimental psychology. He was a founder of psychophysics, and he was particularly credited for discovering the nonlinear relationship between psychological sensation and the physical intensity of a stimulus. However, for the purpose of this historical note, his contribution to electrochemistry must be emphasized.

Fechner extensively performed electrochemical experiments, particularly testing Ohm's law for electrolyte solutions in galvanic cells. He was the first to introduce into electrochemical science the parameter of the charge transfer resistance at an electrode/electrolyte interface. This factor is highly important for electrochemical kinetics in modern electrochemistry. Fechner was the first who reported in 1828 the observation of electrochemical oscillations during the anodic dissolution of nickel in nitric acid (10.2A). From that time, numerous systems with nonstationary oscillating behavior have been discovered.

Fechner designed an electroscope (electrometer) with improved sensitivity. This instrument consisted of 800 to 1000 pairs of metal foils, charging two metal condenser plates. A thin strip of gold foil was hung between these plates (10.2B). A deviation of the gold foil from its vertical position was used for sensing an electric charge on it with very high sensitivity.

Fechner was very influential by translating and completely rewriting volume 3 (*Lehrbuch des Galvanismus und der Elektrochemie*) of Jean Baptiste Biot's textbook on experimental physics in which he gives a full account of the state of knowledge in electrochemistry of his time, including his findings.

10.1. Gustav Theodor Fechner (a photograph)

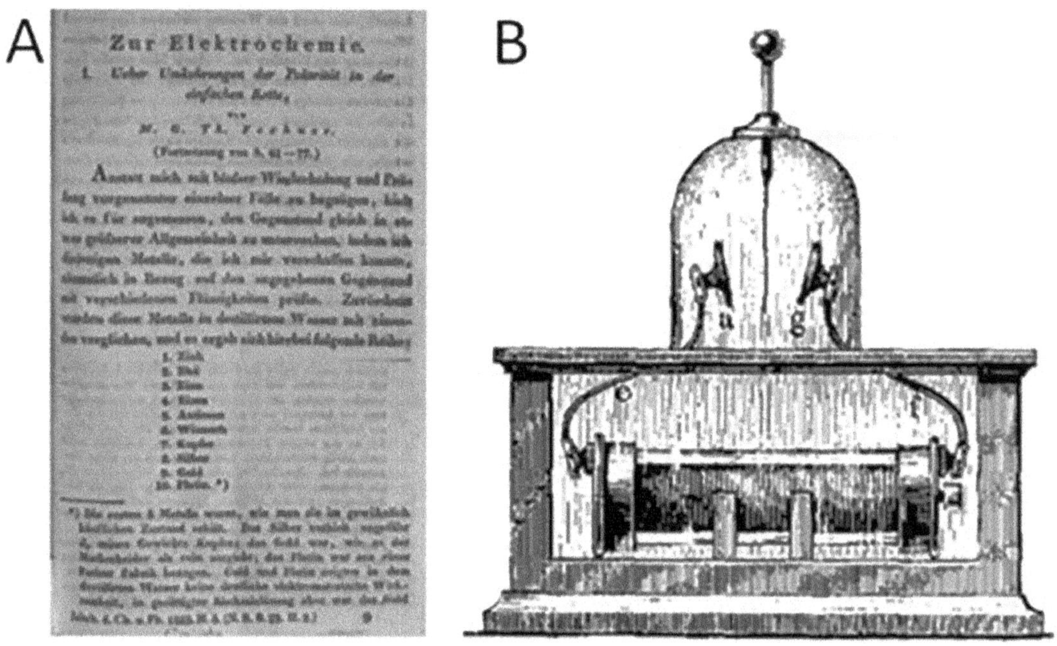

10.2. (A) The first paper by Fechner reporting on oscillating electrochemical systems. (B) Fechner's electroscope.

11

SIR WILLIAM ROBERT GROVE

Born: July 11, 1811, Swansea, United Kingdom
Died: August 1, 1896, London, United Kingdom

William Robert Grove (11.1) was a British scientist who developed a gas voltaic battery, which was the forerunner of modern fuel cells. Thus, Grove is known as the father of the fuel cell.

In his early study, Grove invented a novel electric cell (battery), which was named after him the Grove cell. This new kind of battery included zinc and platinum electrodes (operating as the anode and cathode, respectively) immersed in an acidic solution and separated with a porous ceramic membrane. This battery was demonstrated in 1839 at the *Académie des Sciences* meeting in Paris.

Later in 1842, Grove invented the first fuel cell (named by him "gas voltaic battery"). This cell produced electrical energy by combining hydrogen and oxygen to water at separated electrodes in the process opposite to the water electrolysis. The first fuel cell prototype opened a new research and engineering area leading to modern fuel cells used in many practically important applications. It is interesting to note that the practical importance of fuel cells was not recognized at the beginning. Particularly, the Nobel Prize winner Wilhelm Ostwald (German chemist, 1853–1932) described Grove's gas battery in his famous book *Electrochemistry: History and Theory*, which was published in 1896, as "of no practical importance but quite significant for its theoretical interest." The practical importance of the fuel cells was recognized later. Acknowledging his scientific achievements Grove was knighted in 1872.

11.1. Sir William Robert Grove

12

ADOLPH WILHELM HERMANN KOLBE

Born: September 27, 1818, Elliehausen, Göttingen, Germany
Died: November 25, 1884, Leipzig, Germany

Hermann Kolbe (12.1) was a German scientist who greatly contributed to the development of organic chemistry, transforming it into the state we know it now. Kolbe pioneered organic synthesis from inorganic sources and introduced the term *synthesis* in the meaning how we use it in chemistry now. His name is associated with several synthetic reactions in organic chemistry—e.g., the Kolbe-Schmitt reaction in preparation of aspirin, the Kolbe nitrile synthesis, etc. His work is particularly remembered in connection to the electrolysis of carboxylic acids, resulting in the synthesis of various organic compounds known as the Kolbe reaction.

Kolbe reaction (12.2), proceeding as the electrolysis, results in the oxidative decarboxylation of carboxylic acids, yielding free radicals that dimerize to produce symmetrical products. For example, the Kolbe electrolysis process can proceed in an aqueous solution of sodium acetate (12.2A). The acetate ions get decomposed and form methyl radicals. These combine with other free methyl radicals, which leads to the generation of ethane.

In general, Kolbe's electrolysis method uses sodium salts of fatty acids to form the corresponding alkanes as products. A similar electrochemical synthesis can be used to produce more sophisticated products (12.2B). If the initial mixture includes two different acids, the reaction results in three different products from the cross-reaction of two different free radicals (12.2C). The Kolbe electrolytic decarboxylation of 1,2-dicarboxylic acids results in the formation of double or triple chemical bonds (12.3). When carboxylic groups are located at a longer distance in a molecule, the electrolytic decarboxylation may result in the intramolecular radical cyclization of the reaction product.

It should be noted that the Kolbe electrolysis reaction may result in the formation of numerous byproducts (12.4). The formation of side products depends on the ease of the follow-up oxidation, which leads to carbenium ions and their subsequent rearrangements. The exact mechanism and kinetics study of the electrochemical Kolbe process have been investigated confirming the complexity of the electrochemical reaction.

12.1. Adolph Wilhelm Hermann Kolbe (ca. 1865)

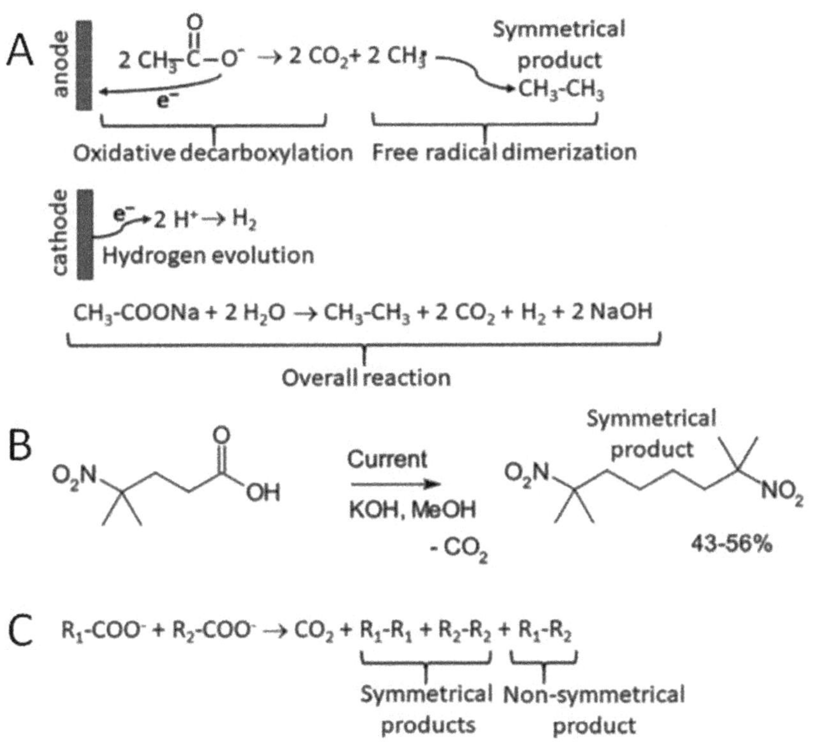

12.2. (A) Kolbe reaction exemplified with electrolysis of acetic acid. (B) More sophisticated example of the Kolbe reaction. (C) Electrolysis of a mixture of two carboxylic acids resulting in three products of the radical cross-reaction

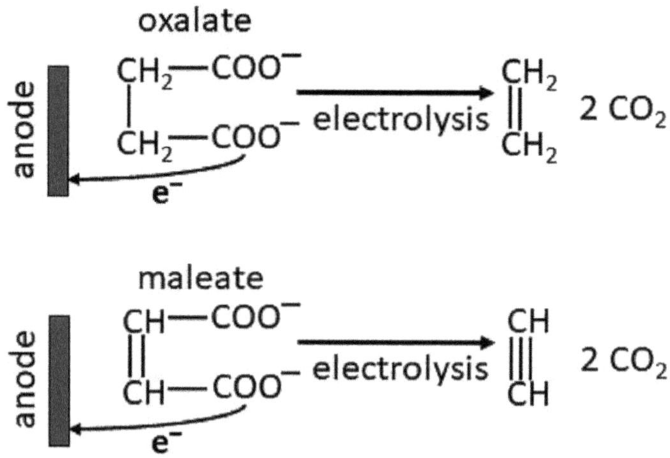

12.3. Kolbe electrolytic decarboxylation of 1,2-dicarboxylic acids results in the formation of double or triple chemical bonds.

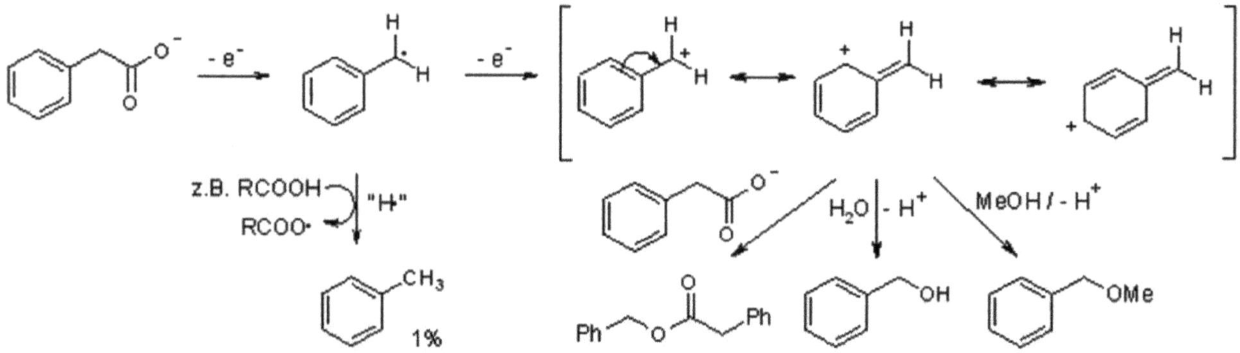

12.4. Formation of by-products in Kolbe electrolysis reaction

13

OLIVER WOLCOTT GIBBS

Born: February 21, 1822, New York City, New York, USA
Died: December 9, 1908, Newport, Rhode Island, USA

Oliver Wolcott Gibbs (13.1) was an American chemist. He is known for performing the first electrogravimetric analyses (13.2), namely based on the reductions of copper (Cu^{2+}) and nickel (Ni^{2+}) ions to their respective metals (Cu^0) and (Ni^0).

In electrogravimetric analysis, the analyte (usually represented by metal cations) is electrochemically deposited (electroplated) on an electrode under potential-controlled conditions. The increase in the weight of the electrode is a direct measure of the amount of metal-analyte in the sample, assuming that it is fully deposited on the electrode. The method is applicable to the systems containing several metal cations in the analyte sample. However, in order to have selective analysis of the metals in the solution sample, the metals in the sample must have significantly different redox potentials to avoid their co-deposition.

When a current is passed through a solution containing two or more electroactive species (different metal cations), the electrochemical process with the most positive reduction potential will occur first at the cathode. For example, if the solution contains copper (Cu^{2+}) and cadmium (Cd^{2+}) ions, copper will deposit first. (Note the redox potentials of these two metals: $E°(Cu^{2+}/Cu^0) = 0.34$ V; $E°(Cd^{2+}/Cd^0) = -0.40$ V.)

When the metallic copper (Cu^0) deposition is completed, the electrode potential is moving to more negative value then resulting in hydrogen gas (H_2) evolution (this is actually water reduction). This step does not change the electrode weight, and this process does not allow the electrode potential to move to more negative values. Overall, the electrochemical reduction of Cd^{2+} cations is not possible as long as the experiment is performed with an aqueous solution. Thus, the Cu^{2+} cations can be separated from the Cd^{2+} cations in the present example, using the electrodeposition technique. In order to optimize the electrodeposition process, different pH values (controlling the $2H^+/H_2$ potential) and different electrode materials (affecting the overpotentials for metal deposition) can be applied.

Finally, the electrode weight before and after the metal electroplating deposition was measured, indicating the metal amount in the analyzed sample. The electrogravimetry technique was successfully applied to the analysis of nickel, copper, cobalt, lead, zinc, manganese, cadmium, and other metals. While the first examples of this analytical technique were studied by Gibbs, many other systems were reported by other scientists in the second half of the nineteenth century.

13.1. Oliver Wolcott Gibbs, 1914

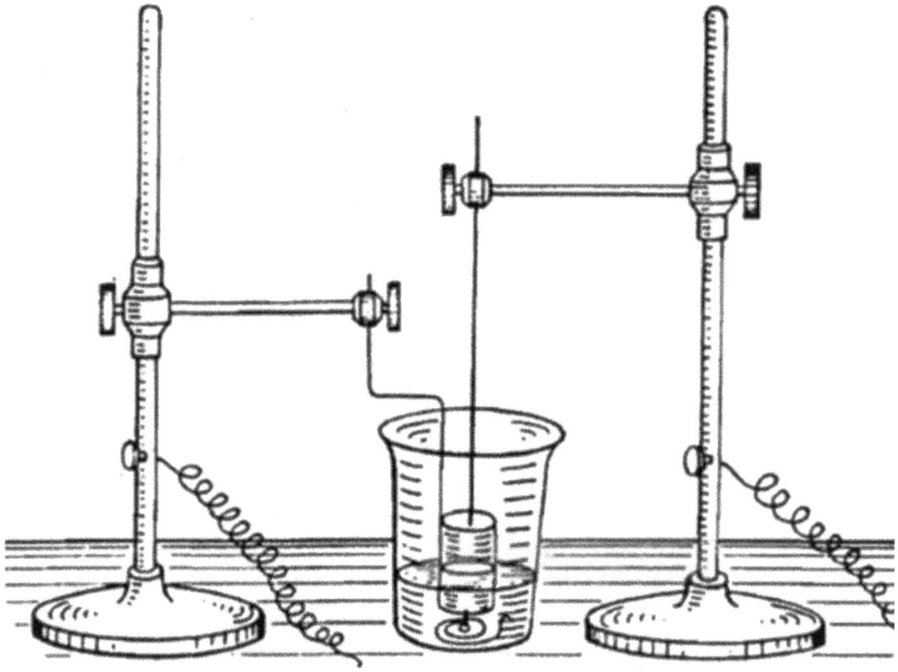

13.2. The electrochemical setup for the electrodeposition of metals followed by their gravimetric analysis

14

JOHANN WILHELM HITTORF

Born: March 27, 1824, Bonn, Germany
Died: November 28, 1914, Münster, Germany

Johann Wilhelm Hittorf (14.1) was a German physicist, the first to find the electrical current carried by oppositely charged ions in electrolyte solutions. He formulated ion transport numbers and the first method for their measurements. The ionic transport is an important factor in understanding electrochemical reactions.

Hittorf was the first to formulate ionic transport theory. He derived theoretically and measured experimentally separate migration of oppositely charged ions to anode and cathode compartments and formulated a method for the measurement of the fraction of the electrical charge carried by the different migrating ions. The ion transport numbers (also called the transference numbers) found by Hittorf are highly important parameters in electrolyte solutions.

In 1853, Hittorf pointed out that some ions traveled more rapidly than others, producing a larger contribution to the ionic current in electrolyte solutions. Hittorf's work on ion transport was mainly made between 1853 and 1859 but remained an important area of study for him for over fifty years. It should be noted that the work, theoretical and experimental, was performed by Hittorf before Svante August Arrhenius (Swedish scientist, 1859–1927), who formulated the theory of ionic dissociation in its final version in his dissertation in 1884.

Besides his work on ionic transport, Hittorf studied allotropic forms of selenium and phosphorus and made significant contribution to the study of cathode rays (X-ray in modern terminology).

14.1. Johann Wilhelm Hittorf (1904)

15

GASTON PLANTÉ

Born: April 22, 1834, Orthy, France
Died: May 21, 1889, Paris, France

Gaston Planté (15.1) was a French physicist who constructed the first practically applicable secondary (rechargeable) battery.

Planté invented the lead-acid battery in 1859 (15.2). This type of battery was developed as the first rechargeable electric battery marketed for commercial use, and it is still widely used in automobiles.

The early model designed by Planté consisted of a spiral roll of two sheets of pure lead, separated by a linen cloth and immersed in a glass jar containing 10 percent sulfuric acid solution. A year later, he presented a battery to the French Academy of Sciences consisting of nine of the elements, housed in a protective box with the terminals connected in parallel. The Planté battery could deliver remarkably large currents.

The redox reactions in the Planté battery are the following:

- $Pb(s) + HSO_4^-(aq) \rightarrow PbSO_4(s) + H^+(aq) + 2e^-$ (negative electrode)
- $PbO_2(s) + HSO_4^-(aq) + 3H^+(aq) + 2e^- \rightarrow PbSO_4(s) + 2H_2O$ (positive electrode)

The voltage produced by the battery was $E^0_{cell} = 2.05$ V. The redox reactions in the battery are reversible, and after discharging, the reversed reactions can be used to recharge the battery again.

15.1. Gaston Planté

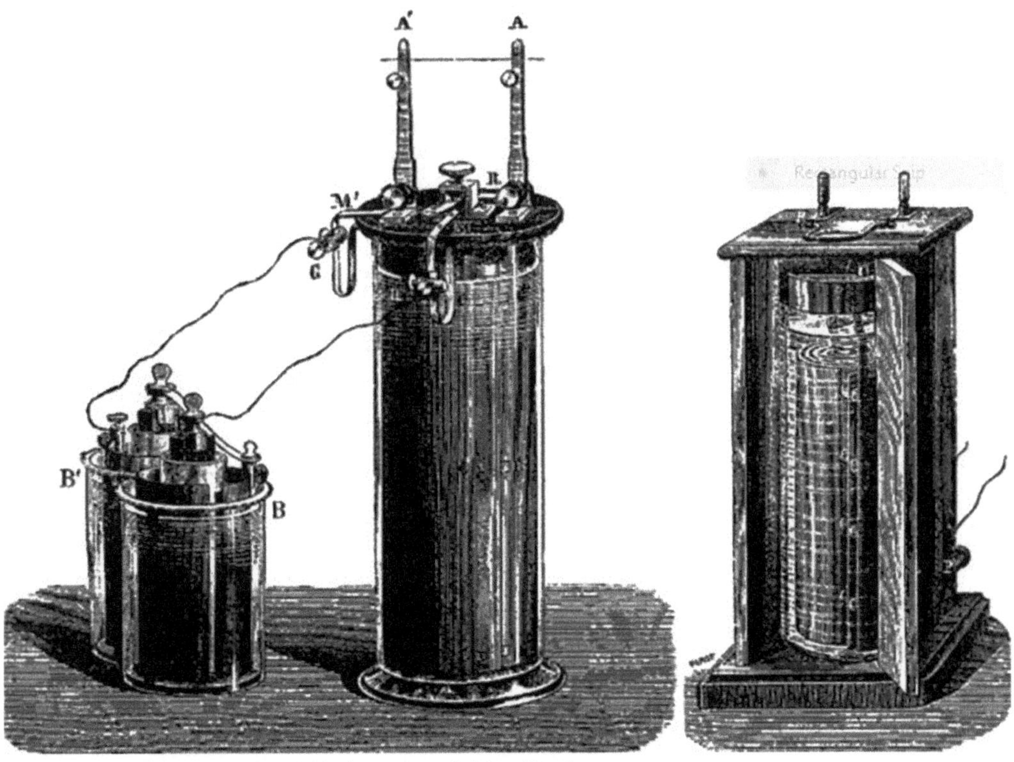

15.2. Early Gaston Planté's lead-acid batteries

16

LUDWIG MOND

Born: March 7, 1839, Kassel, Germany
Died: December 11, 1909, London, United Kingdom

The general concept of fuel cells starts from the experiments of British physicist William Grove (1811–1896), who published the first results on fuel cells in 1839. He used hydrogen and oxygen as a fuel and oxidizer respectively, reacting on platinum catalytic electrodes and generating electric power. However, his research was considered only as scientific proof of the process reversed to the water electrolysis with no practical importance. Indeed, the cell invented by Grove produced a very small current and voltage over a short time. Obviously, after the concept demonstration, some engineering had to be done for improving the cell efficiency to make it feasible for practical use.

During the late 1880s, two British chemists, Ludwig Mond and his assistant Carl Langer (16.1), developed a fuel cell with a longer service life with improved geometry of the catalytic electrodes and flow channels (16.2). They used the known scientific concept from Grove's cell but with improved engineering. Their fuel cell generated 6 amps per square foot current density and 730 mV voltage. The cell operated with coal-derived gas as a fuel and air (actually oxygen in the air) as an oxidizer. The cell was filled with diluted sulfuric acid and included thin perforated platinum electrodes separated with a porous nonconducting membrane. The first engineered fuel cell was demonstrated and patented in 1889. Note that Ludwig Mond and Carl Langer were the first to introduce the term *fuel cell*, which is commonly used now.

16.1. Ludwig Mond (*left*) and Carl (Charles) Langer (*right*)

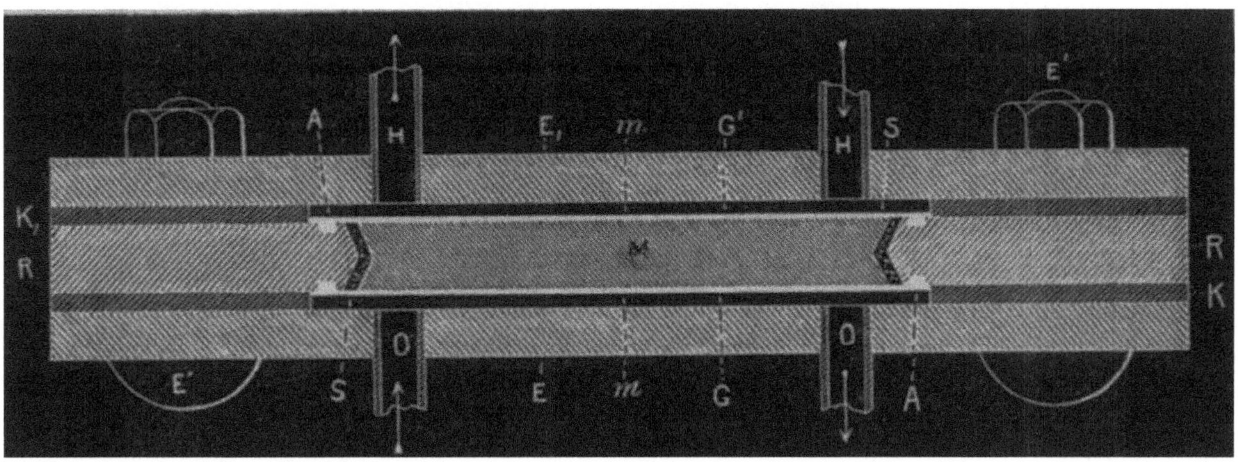

16.2. Mond and Langer's fuel cell design from 1889

17

FRIEDRICH WILHELM GEORG KOHLRAUSCH

Born: October 14, 1840, Rinteln, Germany
Died: January 17, 1910, Marburg, Germany

Friedrich Wilhelm Georg Kohlrausch (17.1) was a German physicist who studied the electrical conductivity of electrolyte solutions and greatly contributed to the understanding of their electrical features, particularly important for electrochemistry and dissociation theory. Kohlrausch's research was concentrated on the theoretical and experimental analysis of electrical conductivity in electrolyte solutions.

Prior to his work, Ohm's law of electric conductivity (reciprocal of electrical resistivity) was mostly applied to metallic materials with electronic conductivity. Kohlrausch attempted to find experimentally and then explain theoretically whether Ohm's law is also applicable to electrolyte solutions with ionic conductivity. Prior to Kohlrausch's work, the conductivity measurements in electrolyte solutions were always complicated by electrolysis of the dissolved species, resulting in the deposition of redox products on electrodes and changing concentrations of the species in the solution. Kohlrausch solved this problem by applying an alternating current for the conductivity measurements, thus keeping the solution composition unchanged. This kind of electric measurement in electrolyte solutions was a precursor of impedance measurements.

Another way of conductivity measurements in electrolyte solutions was based on the use of an electric bridge, which compensated currents and compared the electrolyte resistance with standard resistances in the bridge (17.2). When the electrolyte resistance was equal to the resistance in the bridge, the current was not passing through the cell and the electrolyte composition was not changed. This allowed highly precise results in Kohlrausch's experiments. It is interesting to note that the resistance balancing the bridge, when the current through the bridge was zero, was detected by Kohlrausch using a telephone receiver.

From 1875 to 1879, Kohlrausch performed numerous experiments with aqueous solutions of different salts, acids, and other electrolytes. His experiments confirmed that Ohm's law is also applicable to electrolyte solutions. The theoretical treatment of the experimental data resulted in the conclusion that each type of migrating ions has a specific conductivity, and it does not depend on the origin of the ions. In other words, Na^+ cations have the same contribution to the total conductivity regardless of their origin from NaCl or Na_2SO_4 salts, provided Na^+ concentration is the same.

This experimentally confirmed conclusion resulted in the law of the independent migration of ions, formulated by Kohlrausch. According to this law, the total conductivity of electrolyte solutions is a sum of individual conductivities of all migrating ions (cations and anions) present in the solution. In a simple way, Kohlrausch's conclusions were applicable to strong electrolytes, which are fully dissociated forming ions (except in very high concentrations). More complex behavior was observed for weak electrolytes, which decreased their dissociation with the increased concentrations.

Kohlrausch's law on the independence of migrating ions. The molar conductivity of an electrolyte (Λ) equals the sum of the molar conductivities of the cations (Λ^+) and the anions (Λ^-), taking into account stoichiometric coefficients of the cations (ν^+) and anions (ν^-). See equation 1.

$$\Lambda = \nu^+ \Lambda^+ + \nu^- \Lambda^- \qquad (1)$$

Kohlrausch's square root law. For strong electrolytes with sufficient dilution, the molar conductivity (Λ_c) of an electrolyte of concentration C is a linear function of the root of the concentration (equation 2), where Λ_∞ is a molar conductivity at infinite dilution and A is a constant which is a function of the valence of the salt.

$$\Lambda_c = \Lambda_\infty - A\sqrt{C} \qquad (2)$$

Kohlrausch studied water dissociation and described this process in terms similar to the presently used. The process was explained as hydrogen ion (proton) transfer from one water molecule to another one, thus resulting in the formation of hydroxide ion (OH^-) and hydroxonium ion (H_3O^+). Kohlrausch performed careful purification of water using nearly a hundred vacuum rectifications to avoid any other ionic species in the water. Then based on the conductivity measurements, he obtained an important number of the ionic product for water $[H_3O^+] \times [OH^-] = 1\times10^{-14}\ M^2$ at 25°C. Kohlrausch's study of water dissociation created the background for the formulation of pH as a logarithmic value of the H_3O^+ concentration (equation 3) and the pH scale for aqueous solutions, both highly important for chemical reactions in aqueous solutions in general and for electrochemistry particularly.

$$pH = -\log [H_3O^+] \qquad (3)$$

Famous scientists—particularly those who contributed to the theory of ionic dissociation and theoretical chemistry, including Svante Arrhenius (Swedish scientist, 1859–1927); Wilhelm Ostwald (Baltic German chemist, 1853–1932); and Jacobus Henricus van't Hoff (Dutch physical chemist, 1852–1911)—grew inspired by the work of Kohlrausch (17.3). Kohlrausch greatly contributed to education in physics, particularly with his textbooks. Kohlrausch's book *Leitfaden der praktischen Physik* (*Guidelines to Practical Physics*) was published in 1870 and was reprinted many times under the title *Praktische Physik* (*Practical Physics*). It was used over many years as a classical textbook. It contained not only descriptions of experiments, experimental setups, and measuring techniques but also tables of physical quantities.

17.1. Friedrich Wilhelm Georg Kohlrausch, ca. 1875 (portrait by Otto Patzig, a German portrait painter and lithographer, 1822–1885)

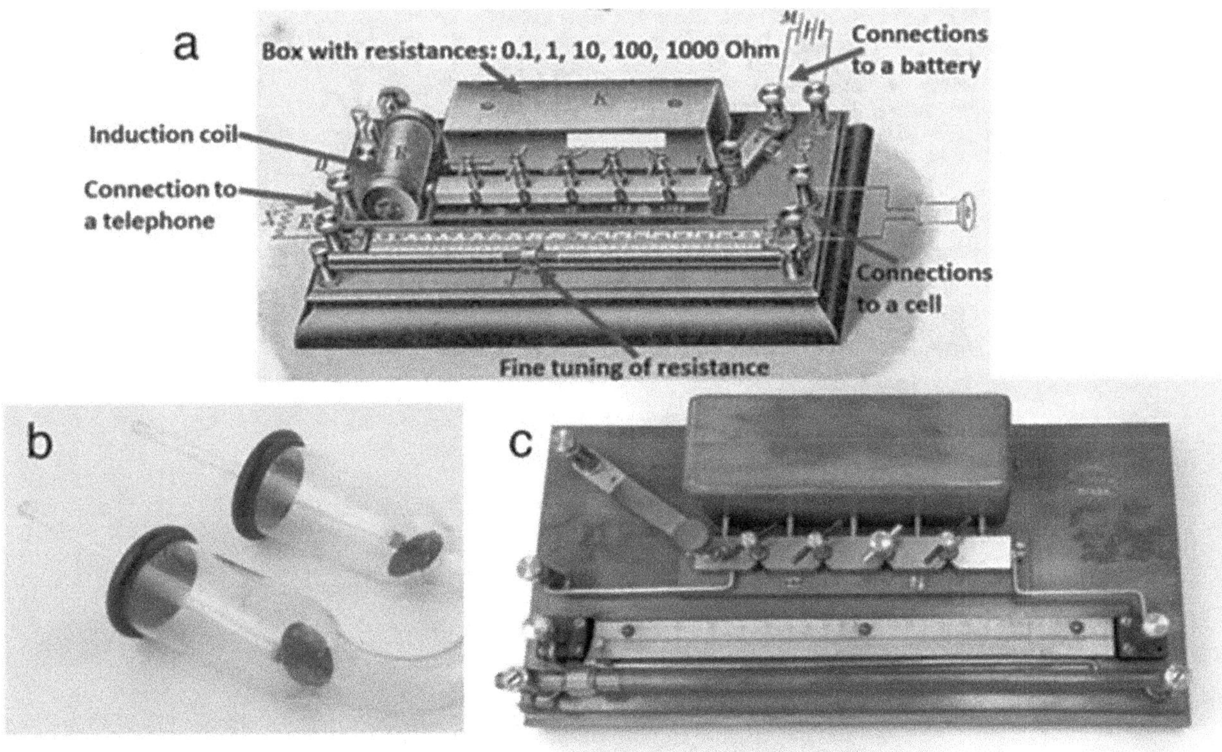

17.2. Kohlrausch's instruments used for conductivity measurements in electrolyte solutions: (a) Kohlrausch's electrical bridge—schematics; (b) Kohlrausch's cell with nickel electrodes; (c) Kohlrausch's electrical bridge (produced by Hartmann & Braun, Frankfurt a. Main in 1895; this version does not include an induction coil)

17.3. Friedrich Kohlrausch with his colleagues, including famous scientists Svante Arrhenius and Walther Nernst (photo in Physical Institute, University of Wurzburg, winter semester 1886–1887): *Standing from the left*: Heydweiller, Rasch, Arrhenius, Nernst. *Sitting from the left*: Palazzo, Kohlrausch, Sheldon.

18

EMIL GABRIEL WARBURG

Born: March 9, 1846, Altona, Duchy of Holstein (presently Germany)
Died: July 28, 1931, Bayreuth, Weimar Republic (presently Germany)

Emil Gabriel Warburg (18.1) was a German physicist. His name is notably associated with the Warburg element of electrochemical impedance.

Warburg was the first to extend the concept of impedance from electrical/electronic systems to electrochemical systems at the beginning of the nineteenth century when he derived the impedance function for a diffusional process.

The impedance of electrochemical systems is a complex function of many simultaneous processes, including resistance of the interfacial electron transfer, bulk solution resistance, interfacial capacitance, diffusion impedance, and more. In order to study this sophisticated function, it can be represented with a simplified model, which includes various electrical elements (capacitances, resistances) organized in a special model circuitry. One of the elements in this model is the Warburg diffusion element that models the diffusion process in impedance spectroscopy. When analyzing an experimental impedance spectrum, a Warburg impedance element is not easy to separate from a charge-transfer resistance (in the bulk) and a double layer capacitance (at the interface). The presence of the Warburg element in the impedance spectrum can be found if a linear relationship on the log of a Bode plot (log $|Z|$ vs. log ω) exists with a slope of value $-1/2$ (where Z is the experimental impedance, ω is the applied angular frequency).

The Warburg diffusion element (Z_W) is a constant phase element (CPE), with a constant phase of 45° (phase independent of frequency) and with a magnitude inversely proportional to the square root of the applied frequency. The Warburg frequency dependence is expressed by the following equations:

$$Z_W = \frac{A_W}{\sqrt{\omega}} + \frac{A_W}{j\sqrt{\omega}}$$

$$|Z_W| = \sqrt{2}\frac{A_W}{\sqrt{\omega}}$$

where A_W is the Warburg coefficient (or Warburg constant); j is the imaginary unit; ω is the angular frequency. These equations assume unrestricted diffusion to a large planar electrode. Notably, electrodes with different geometry (e.g., microelectrodes) will result in a different kind of diffusion and different frequency dependences.

Overall, the Warburg element represents the contribution of diffusion to the impedance spectra. Actually, it is a virtual process in the electrochemical reaction analyzed with impedance spectroscopy, and it is needed to make a convenient presentation of equivalent circuits for electrochemical experiments.

18.1. Emil Gabriel Warburg

19

FERDINAND FREDERIC HENRI MOISSAN

Born: September 28, 1852, Paris, France
Died: February 20, 1907, Paris, France

Ferdinand Frederic Henri Moissan (19.1) was a French (with Jewish ethnicity) chemist who obtained pure fluorine gas (F_2) by electrolysis of the F-containing chemicals. Notably, he was awarded with a Nobel Prize in Chemistry (1906) for this discovery.

Fluorine-containing materials have been known and used for many centuries. The first known use of a fluorine compound dates to around 1670 being a set of instructions for etching glass with the compound called Bohemian emerald (CaF_2). However, because of the high chemical activity of fluorine, it was very difficult to produce the elemental fluorine. The F^- anions cannot be chemically oxidized to produce elemental F^0 (actually F_2) material. Indeed, the highest electronegativity of the F atoms does not allow chemical oxidation of F^- anions. Electrochemical oxidation of F^- anions was the only chance for producing elemental fluorine. Unfortunately, the electrolysis in aqueous solutions resulted only in the production of oxygen (O_2) because of oxidation of water.

In 1886, Moissan was the first to produce pure elementary fluorine by electrolysis of KF salt dissolved in liquid anhydrous HF acid (19.2). The electrolysis was performed using platinum electrodes in a cell under low temperature (-23°C). Note that at room temperature HF is a gas. The solution must be anhydrous; otherwise, the electrolysis would produce oxygen gas (O_2) at the anode instead of the F_2 gas. Indeed, the water oxidation resulting in the O_2 formation proceeds easier (at less positive potentials) than F^- ion oxidation.

With that experiment, Moissan proved that electrolysis was capable of oxidizing species such as fluoride ions for which no chemical oxidant was strong enough. Electrolysis is still used today to produce fluorine on an industrial scale; however, KF is now electrolyzed as a molten salt at 250°C or molten KH_2F_3 salt used at 50°C.

Another result from Moissan work was the construction of the electric arc furnace in 1892. The furnace was attaining temperatures up to 3,500°C. This was mostly technical but still very important achievement, which allowed him many chemical experiments at high temperature. In the furnace, Moissan managed to synthesize microscopic diamonds (according to his scientific reports) and other materials. According to later reports by other scientists, Moissan did not produce diamonds, but the produced material was silicon carbide, later named *carborundum*, which has a crystal structure like that of diamond and is almost as hard.

Moissan's work on producing F_2 and construction of the electric arc furnace was well accepted by the scientific community and resulted in the Nobel Prize. The Nobel Prize in Chemistry 1906 was awarded to Henri Moissan "in recognition of the great services rendered by him in his investigation and isolation of the element fluorine, and for the adoption in the service of science of the electric furnace called after him."

19.1. Ferdinand Frederic Henri Moissan

19.2. Moissan's fluorine cell, from his 1887 publication (*left*): *F* corresponds to the anodic production of F_2 gas; *H* corresponds to the cathodic production of H_2 gas. The cube below the electrolytic cell is ice for maintaining the low temperature. The wires are connected to an electric battery. Moissan shown in his lab isolating fluorine, 1886 (*right*); note the same electrolytic cell as shown at left.

20

LOUIS GEORGES GOUY

Born: February 19, 1854, Vals-des-Bains, Ardeche, France
Died: January 27, 1926, Vals, Ardeche, France

Louis Georges Gouy (20.1) was a French physicist who was particularly remembered in electrochemistry for the theory of the diffuse electrical double layer formed at an electrode interface upon polarization of an electrode (20.2).

Gouy derived the equations, describing his double-layer model for 1:1 electrolytes (e.g., Na^+/ Cl^-) and for asymmetrical electrolytes (e.g., Ca^{2+} / 2 Cl^-). The width of the diffuse double layer at an electrode depends on a number of factors among which the charge density on the surface of the electrode and the concentration of the electrolyte in the solution are important.

David Leonard Chapman (English physical chemist, 1869–1958) performed a study related to Gouy's electrolyte model, then his name is usually mentioned together with Gouy as the Gouy-Chapman electric double layer model. This model, despite its simplicity and limitations, describes the electrical double layer and finds applications in many areas of studies from physical chemistry to biophysics.

Gouy also published several papers on the electrocapillary phenomenon (later important for polarography). His famous work on Brownian motion was important for colloid science. Gouy worked also on optical interference and magnetic effects.

Gouy's name is mentioned in several devices, theoretical models, and processes—e.g., the Gouy balance and the Gouy phase. A Gouy balance (20.3) is used to determine the magnetic susceptibility of a substrate. A strong electromagnet is placed next to the sample, which is on a balance. If the sample is paramagnetic, the mass reading of the balance will increase when the magnetic field is switched on.

Note that in some books and papers, his name cited erroneously as M. Gouy because many of Gouy's papers refer to him as M. Gouy. However, M. denotes monsieur in French, not an initial.

20.1. Louis Georges Gouy performing an experiment with an electric battery (many electric cells connected together are shown in the left-bottom corner)

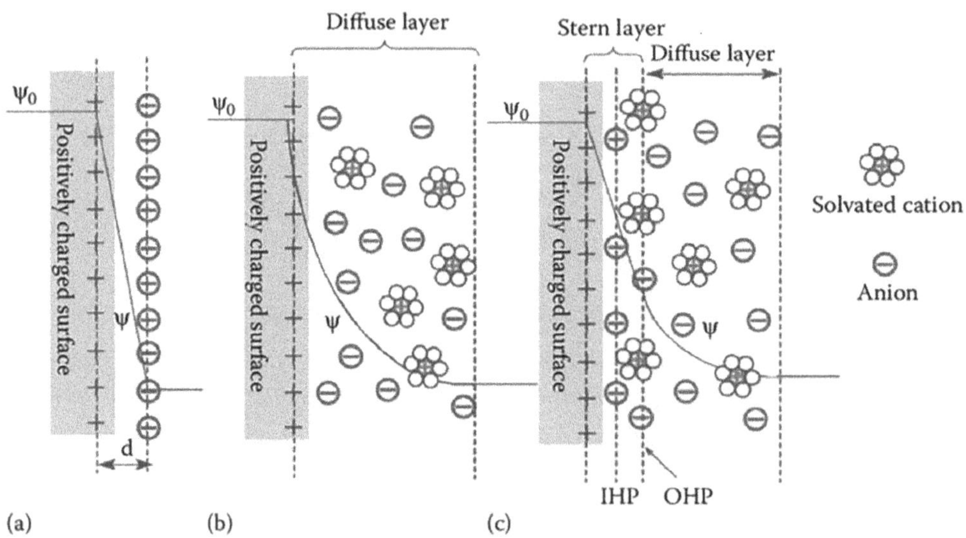

20.2. Three models of the electrical double layer at a positively charged surface: (a) the Helmholtz model, (b) the Gouy-Chapman model, and (c) the Stern model. The Gouy-Chapman model corresponds to the diffuse layer of ions near the electrode interface. The Helmholtz model is a simplified description of the double layer forming a counter ion-layer strongly associated with the electrode surface. The Stern model is a combination of the Helmholtz and Gouy-Chapman models.

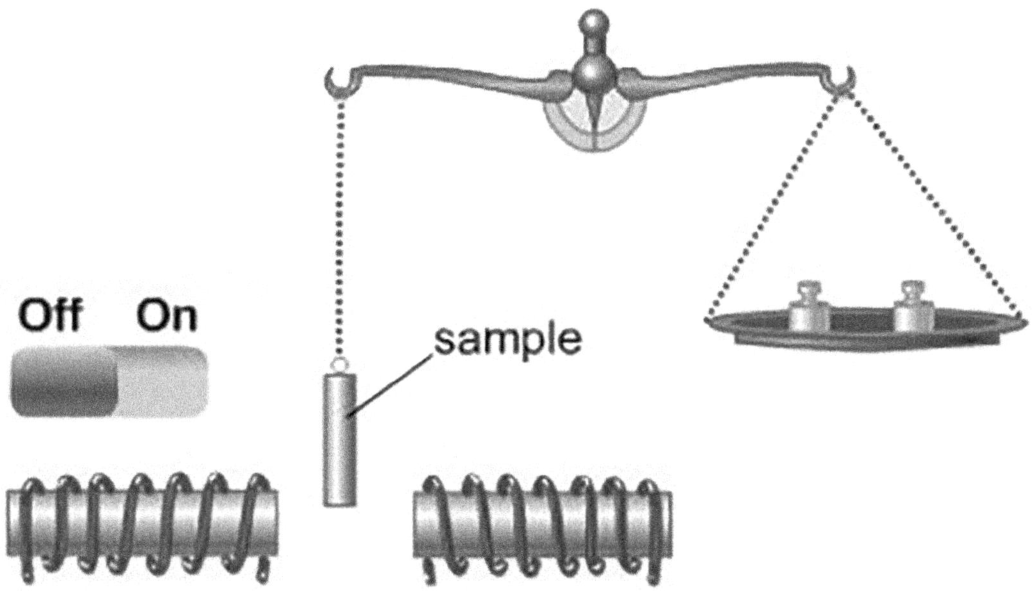

20.3. Gouy balance shown schematically

21

WILLIAM W. JACQUES

Born: 1855, USA
Died: 1932, USA

William W. Jacques was an American electrical engineer and chemist who designed in 1896 a very unusual fuel cell operated on solid (coal) fuel named by him a carbon battery. While the majority of fuel cells utilize gas (usually H_2) or liquid (e.g., ethanol) fuel, the carbon battery designed by Jacques was based on a carbon electrode operated as the fuel (21.1).

The battery consisted of a hundred cells connected in series and placed on top of a furnace that kept the electrolyte temperature between 400–500°C. The produced electrical output was 16 A at 90 V. Based on the experimental results, Jacques claimed around 82 percent efficiency for his carbon battery, but careful analysis considering the heat energy used in the furnace and the energy used to pump air (O_2 was an oxidizer) resulted in a much lower battery efficiency of around 8 percent.

Later research demonstrated that the current generated by his battery was not obtained through an electrochemical reaction as suggested by Jacques but rather through the thermoelectric effect. Several subsequent researchers have stated that Jacques's work was the last notable attempt to derive electricity directly from coal.

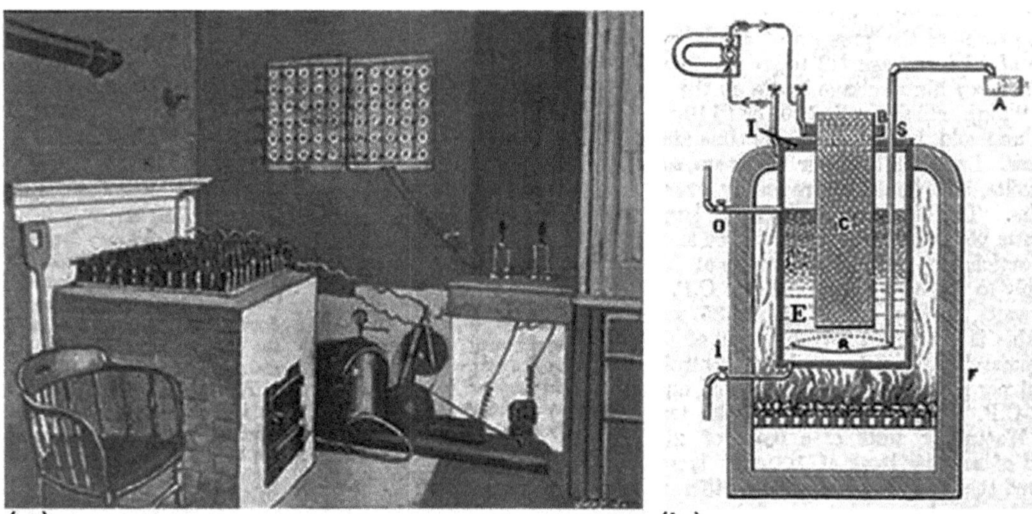

(a) (b)

21.1. (a) The carbon battery designed by Jacques. (b) The furnace used in the battery.

22

ANTON FRIEDRICH ROBERT BEHREND

Born: December 17, 1856, Harburg, Germany
Died: September 15, 1926, Hannover, Germany

Anton Friedrich Robert Behrend (22.1) was a German chemist mostly known for his fundamental research in synthetic organic chemistry, especially related to biologically important materials. Behrend is also well-known for the so-called Behrend rearrangement in organic chemistry. However, in the present book, he should be mentioned for the invention of potentiometric titration.

Behrend performed the first potentiometric titration (with precipitation of the redox species) in the laboratory of Friedrich Wilhelm Ostwald (German scientist, 1853–1932) in Leipzig, Germany. Behrend titrated a mercury (I) nitrate solution with chloride, bromide, and iodide ions using a mercury electrode as an indicator electrode. In the presence of Cl^- or Br^- or I^-, Hg_2^{2+} ions produce insoluble precipitate Hg_2Cl_2 (named calomel) or similar insoluble salts with Br^-, I^-, thus decreasing the concentration of the Hg_2^{2+} ions in the solution:

$$Hg_2^{2+} + 2\ Cl^- \rightarrow Hg_2Cl_2$$

This resulted in the potential change of the Hg droplet indicator electrode. Note that the Hg electrode potential is controlled by the reversible redox reaction: Hg^0/Hg_2^{2+} (similarly to the calomel reference electrode operation).

22.1. Robert Behrend (ca.1925)

23

IDA HENRIETTA HYDE

Born: September 8, 1857, Davenport, Iowa, USA
Died: August 22, 1945, Berkeley, California, USA

Ida Henrietta Hyde (23.1) was an American physiologist known for developing a microelectrode applied for electrical stimulation of biological tissue.

Hyde, in addition to be a university professor and researcher, was an inventor of instrumentation needed in her experimental work. Particularly, she created instruments, which were not available commercially, for measuring physiological parameters in marine animals. The most practically important invention was a micropipette electrode used by her in physiological studies (23.2). The experiments performed by Hyde demonstrated electrically stimulated cell division.

Importantly, the electrical effects were applied at the microlevel of individual (single) cells. The microelectrode made by Hyde (23.3) was a revolutionary invention in neurophysiology and in the study of contractile nerve tissue. Unfortunately, Hyde considered this electrode only as a technical component of her scientific study, and she did not patent or describe it in her publications in detail.

As a result, the microelectrode was not officially attributed to Hyde as being its inventor. While Hyde already used this microelectrode in 1921 (based on some sources already in 1914–1919), microelectrodes with a similar design were reported by other researchers much later but still claimed by other scientists as novel instrumentation. Particularly, Ralph W. Gerard (American neurophysiologist and behavioral scientist, 1900–1974) made a similar microelectrode twenty years later and then was nominated for a Nobel Prize (but did not receive it) for the experiments with his microelectrode.

While the original use of the microelectrode was in biological studies, much broader applications in electroanalytical chemistry emerged later, particularly in scanning electrochemical microscopy. In these studies, the microelectrodes (often named ultramicroelectrodes) have a configuration different from the earlier design developed by Hyde. The novel design of the ultramicroelectrodes was invented in 1980s by Robert Mark Wightman (American electrochemist, 1947) and Martin Fleischmann (British chemist, 1927–2012).

23.1. Ida Henrietta Hyde

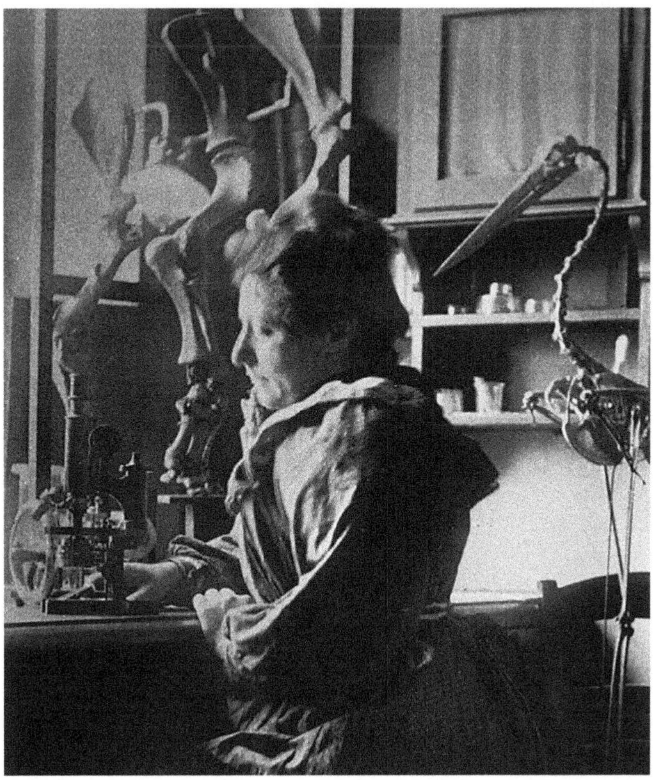

23.2. Ida Henrietta Hyde in her laboratory in Heidelberg, Germany (ca. 1896)

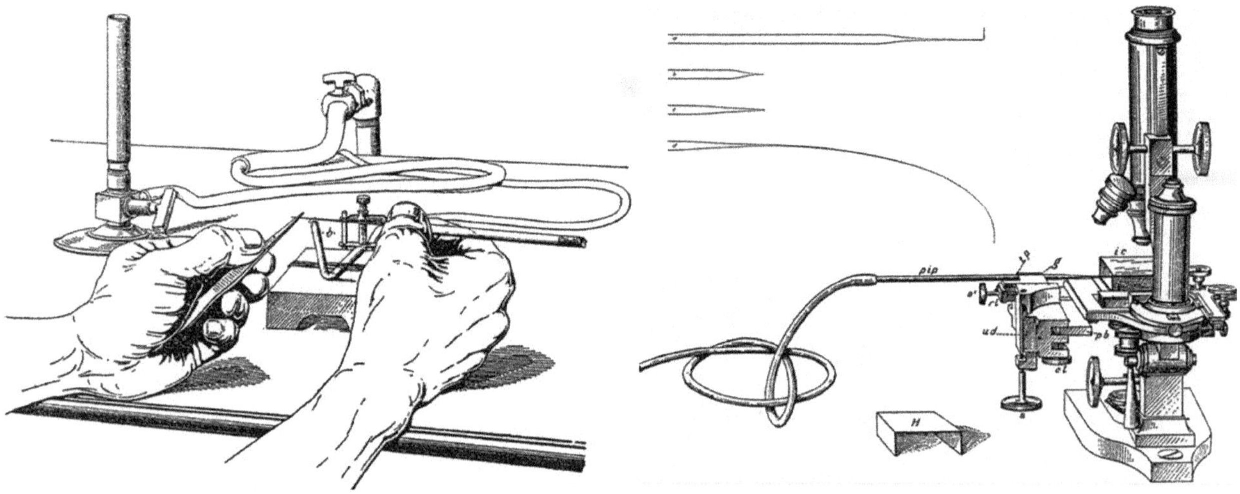

23.3. Preparation of glass micropipettes with a microelectrode inside (1914)

24

SVANTE AUGUST ARRHENIUS

Born: February 19, 1859, Wik, Sweden
Died: October 2, 1927, Stockholm, Sweden

Svante August Arrhenius (24.1) was a Swedish scientist, educated as a physicist, but mostly contributed to chemistry. He established a new scientific field of *physical chemistry*. While he was not the only founder of this novel area combining physics and chemistry, his work was critically important for the formulation and methodology of physical chemistry.

The most important scientific contribution made by Arrhenius was the invention of the electrolytic dissociation theory. This theory explained ionic conductivity in salt/acid/base-solutions and provided background for research of electrochemical processes, including electroanalytical chemistry, electrolysis, and battery chemistry. The first formulation of this theory, presently known as Arrhenius dissociation theory, was made in his PhD thesis submitted in 1884 titled *Recherches sur la Conductibilite Galvanique des Electrolytes* (*Investigations on the Galvanic Conductivity of Electrolytes*).

The theoretical assumption made by him was well supported with extensive experimental work made by Arrhenius, still being a student. The electrical conductivity in aqueous solutions of salts, acids, and bases was explained by splitting the dissolved molecules or crystals in ions (positively charged cations and negatively charged anions). Particularly for acids and bases, he suggested their definitions based on the generation of H^+ and OH^- ions in the case of acids and bases, respectively. This definition of the acids and bases still keeps his name: Arrhenius acids and Arrhenius bases.

The Arrhenius theory had some connections to the early work made by Michael Faraday (English scientist, 1791–1867). Faraday, while studying electrolysis process, also proposed generation of cations and anions supporting conductivity in solutions. However, Faraday believed that their formation proceeds at electrode surfaces only upon pathing electric current through solutions. This explanation is incorrect according to the modern science. The Arrhenius theory proposed the cation and anion formation just upon dissolution of salts, acids, or bases, regardless of the electric current applied. The dissociation of molecules into cations and anions (24.2), according to Arrhenius theory, proceeds due to weakening polar chemical bonds in solutions based on solvents with the high dielectric constants (high polarity of the solvent molecules, water in the original Arrhenius work). This explanation appears to be correct.

Another important contribution of Arrhenius to physical chemistry was his equation describing the temperature dependence of chemical reaction rates proposed in 1889. This equation was based on the experimental work of Jacobus Henricus van't Hoff (Dutch chemist, 1852–1911). Arrhenius provided a physical explanation for the experimentally found empirical formula. The Arrhenius equation provides a relationship between a rate constant of a chemical reaction, k, and absolute temperature, T (Kelvin scale). His equation is given by

$$k = Ae^{\frac{-E_a}{RT}}$$

where A is the pre-exponential factor, E_a is the molar activation energy, R is the universal gas constant.

It is interesting to note that the theory of electrolytic dissociation was so novel that it was poorly accepted by the scientific community; particularly, his PhD thesis received a low score. Notably, later, his theory was awarded the Nobel Prize. Arrhenius received the Nobel Prize for Chemistry in 1903, becoming the first Nobel laureate in Sweden. Shortly after that, in 1905, Arrhenius became the director of the Nobel Institute, where he remained in this important position until his death in 1927. In recognition of his scientific contributions, the Arrhenius dissociation theory, Arrhenius equation, Arrhenius acid, and Arrhenius base include his name.

24.1. Svante Arrhenius (1909), photo made by Meisenbach Riffarth & Co. Leipzig, published in *Zeitschrift für Physikalische Chemie*, vol. 69, 1909

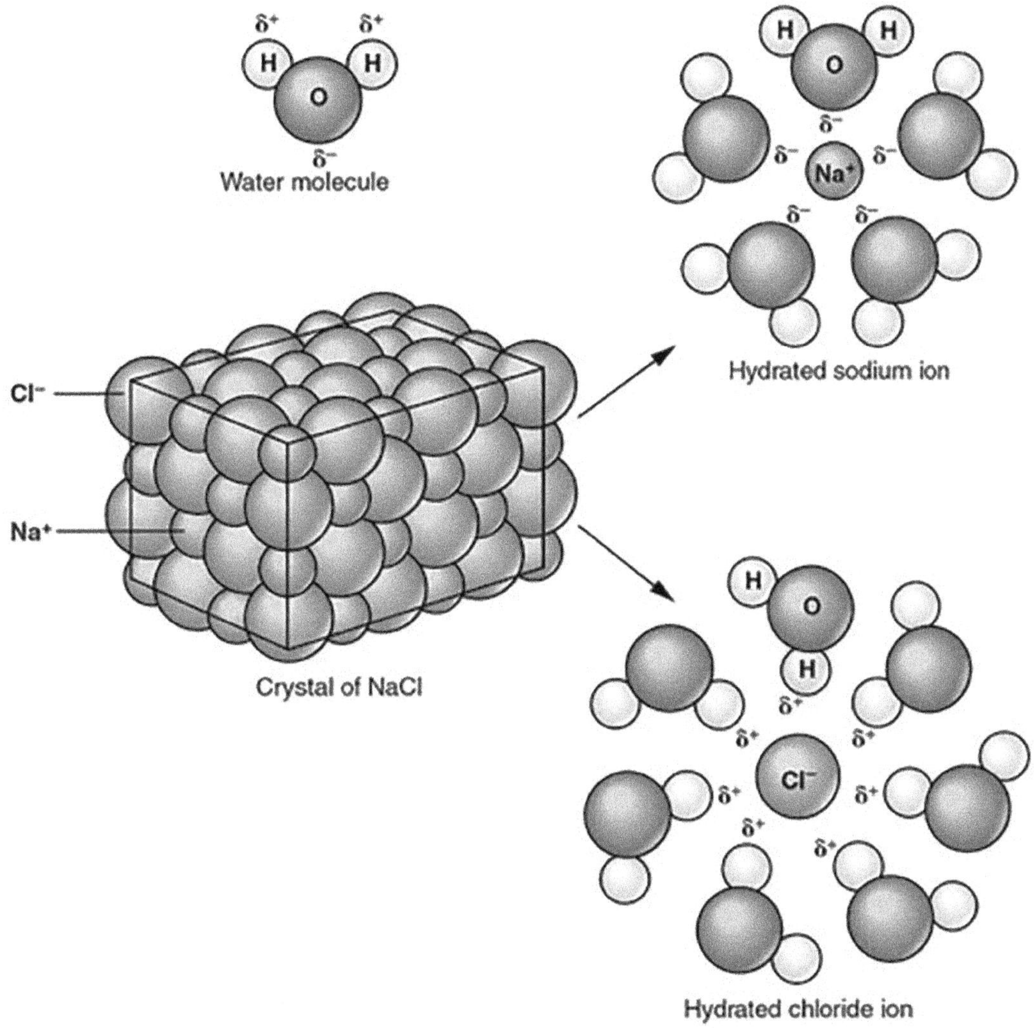

24.2. Schematically shown dissociation of NaCl crystals into hydrated Na^+ cations and Cl^- anions in an aqueous solution

25

JULIUS TAFEL

Born: June 2, 1862, Courrendlin, Switzerland
Died: September 2, 1918, Munich, Germany

Julius Tafel (25.1) was a Swiss chemist and electrochemist. Tafel started his scientific career working in the field of organic chemistry with Hermann Emil Fischer (German chemist, 1852–1919) but soon changed his interests to electrochemistry after his work with Wilhelm Ostwald (German chemist, 1853–1932).

Then Tafel's work concentrated on the electrochemistry of organic compounds and the relation between rates of electrochemical reactions and applied overpotentials. Tafel's name is presently associated with many electrochemical terms: *Tafel equation*, *Tafel slope*, *Tafel rearrangement*, and *Tafel mechanism of hydrogen evolution*.

The Tafel equation and the corresponding Tafel plot (25.2) in electrochemical kinetics relate the rate of an electrochemical reaction in terms of the current density (i) to the overpotential (η) applied. The Tafel equation was first deduced experimentally and was later shown to have a theoretical justification. Indeed, it represents a simplified version of the theoretically derived Butler-Volmer equation (25.2) when the overpotentials are rather high ($|\eta| > 0.1$ V; Tafel region). For a large overpotential (anodic or cathodic), one part of the Butler-Volmer equation becomes negligible while the second part can be transformed into the Tafel equation. The Tafel slope (A) shows how much the overpotential needs to be increased to increase the reaction rate (which is current in electrochemistry) by tenfold.

In a simple case of a one-electron transfer electrochemical reaction, the Tafel slope is determined by the symmetry factors (α_a and α_c), which are usually approximately 0.5, translating to a Tafel slope (A) of 120 mV. The Tafel equation, empirically derived from his experiments with the electrochemical evolution of H_2, lay the background for a new scientific area of electrochemical kinetics. Tafel is also credited for the discovery of the catalytic mechanism of hydrogen evolution (the Tafel mechanism) and construction of a new kind of hydrogen coulometer used in his study of H_2 evolution. Also, he demonstrated that hydrocarbons with isomerized structures can be generated upon electrochemical reduction of the respective acetoacetic esters (named Tafel rearrangement) (25.3). This was an important method for the synthesis of certain hydrocarbons from alkylated ethyl acetoacetate, a reaction accompanied by the rearrangement reaction of the alkyl group.

25.1. Julius Tafel (a photo, ca. 1888)

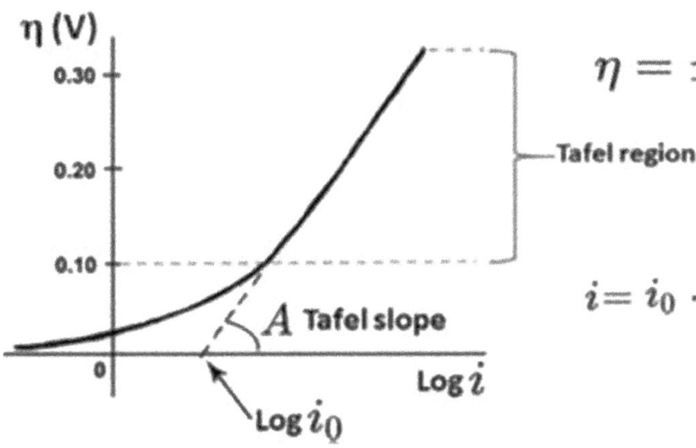

$$\eta = \pm A \times \log_{10}\left(\frac{i}{i_0}\right) \quad \textbf{Tafel equation}$$

Butler–Volmer equation

$$i = i_0 \cdot \left\{\exp\left[\frac{\alpha_a zF\eta}{RT}\right] - \exp\left[-\frac{\alpha_c zF\eta}{RT}\right]\right\}$$

25.2. Tafel plot, Tafel equation, and Butler-Volmer equation. The Tafel region corresponds to the part of the function well approximated with the Tafel equation. In the Tafel equation, η is overpotential (V); ± corresponds to anodic (+) or cathodic (–) polarization; *i* is current density (A/m^2); i_0 is exchange current density (A/m^2) when η = 0; *A* is Tafel slope. In the Butler-Volmer equation, *i*, i_0, and η have the same meaning as in the Tafel equation; α_a and α_c are anodic and cathodic charge transfer coefficients (dimensionless); *F* is Faraday constant (C/mol); *R* is the universal gas constant (J/mol K); *T* is the absolute temperature (K); z is the number of electrons involved in the electrode reaction.

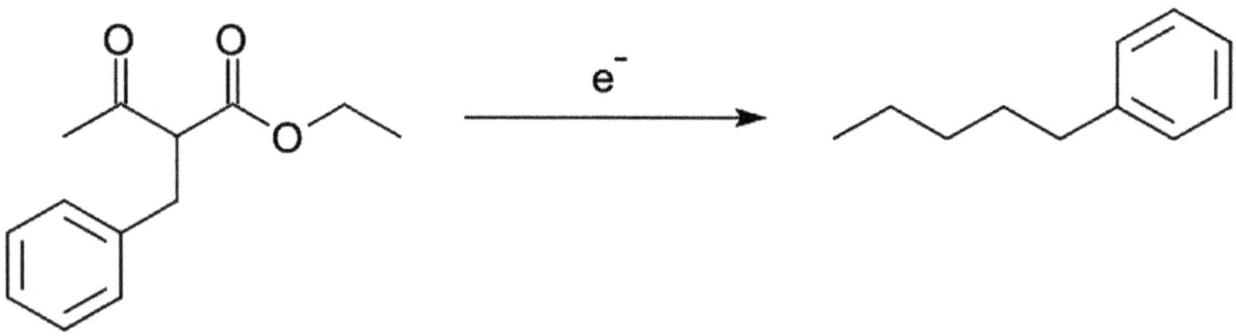

25.3. An example of the Tafel rearrangement electrochemical reaction (by-products are not shown)

26

CHARLES MARTIN HALL

Born: December 6, 1863, Thompson, Ohio, USA
Died: December 27, 1914, Daytona Beach, Florida, USA

Charles Martin Hall (26.1A) was an American inventor, businessman, and chemist. He is best known for his invention in 1886 of an inexpensive method for the electrochemical production of aluminum.

Aluminum was difficult to obtain in its pure metallic state. Its chemical reduction from naturally existing oxide minerals was very difficult and expensive. Soon after its discovery, the price of aluminum exceeded that of gold, thus avoiding its practical application.

Progress in electrochemistry provided hope for the inexpensive production of aluminum by electrolysis of its oxides/salts. However, the electrochemical production of metallic aluminum cannot be performed in aqueous solutions because based on its redox potential, it is not reduced at a cathode since water reduction with molecular hydrogen generation starts before the electrode potential reaches the value required for the aluminum reduction. Therefore, the electrolysis must be done in the molten state of the aluminum minerals, excluding water in the system.

Technically, it was possible, but still, it was not commercially profitable because of the very high temperature required to melt the aluminum minerals. Then physical chemistry suggested a solution: The aluminum minerals (Al_2O_3) must be mixed with something to reduce the melting point (note that mixtures always have lower melting points compared to clean mono-component materials). The practical question was what exactly should be added to Al_2O_3 to decrease the melting point substantially and allow the inexpensive production of metallic aluminum upon electrolysis of the molten mixture. The problem was solved independently and almost simultaneously by two scientists: Charles Martin Hall (in the USA) and Paul Louis Toussaint Héroult (French scientist, 1863–1914).

After numerous unsuccessful experiments with different mixtures started in 1881, Hall found a good component for mixing it with Al_2O_3, then allowing the electrolysis of the molten mixture at a reasonably low temperature of *only* 1000°C. The good addition was cryolite (Na_3AlF_6, sodium hexafluoroaluminate). Pure cryolite has a melting point of around 1009°C. With a small percentage of alumina (Al_2O_3) dissolved in it, its melting point drops to about 1000°C. Besides having a relatively low melting point, cryolite was used as an electrolyte because, among other things, it also dissolves alumina well, conducts electricity, dissociates electrolytically, and also has a lower density than aluminum at the temperatures required for the electrolysis. Then the electrolysis of the molten mixture of (Al_2O_3):(Na_3AlF_6) allowed the production of metallic Al on the cathode. Hall had to fabricate most of his apparatus and prepare his chemicals. He was assisted by his older sister Julia Brainerd Hall (26.1B), who made many contributions to the discovery of the electrolytic means of producing aluminum and to the business success of the aluminium production in the USA. Unfortunately, as usual for that time, her contribution to the discovery was not properly acknowledged. In 1886, Hall submitted the first US patent for the electrochemical production of metallic aluminum (26.2).

The invented Hall-Héroult process (26.3, 26.4) resulted in reducing the price of aluminum by a factor of 200, making it affordable for many practical applications. In 1900, the annual production of aluminum already reached over seven million kilograms. Presently, the aluminum production using the Hall-Héroult technology (26.5) exceeds production of all other metals combined, excluding iron.

It is interesting to note that there is a remarkable coincidence. Both inventors, Hall (in the US) and Héroult (in France) were the same year born; invented aluminum production the same year (Hall applied for the US patent on July 9, 1886, while Héroult obtained a patent in France on April 23, 1886); and died the same year.

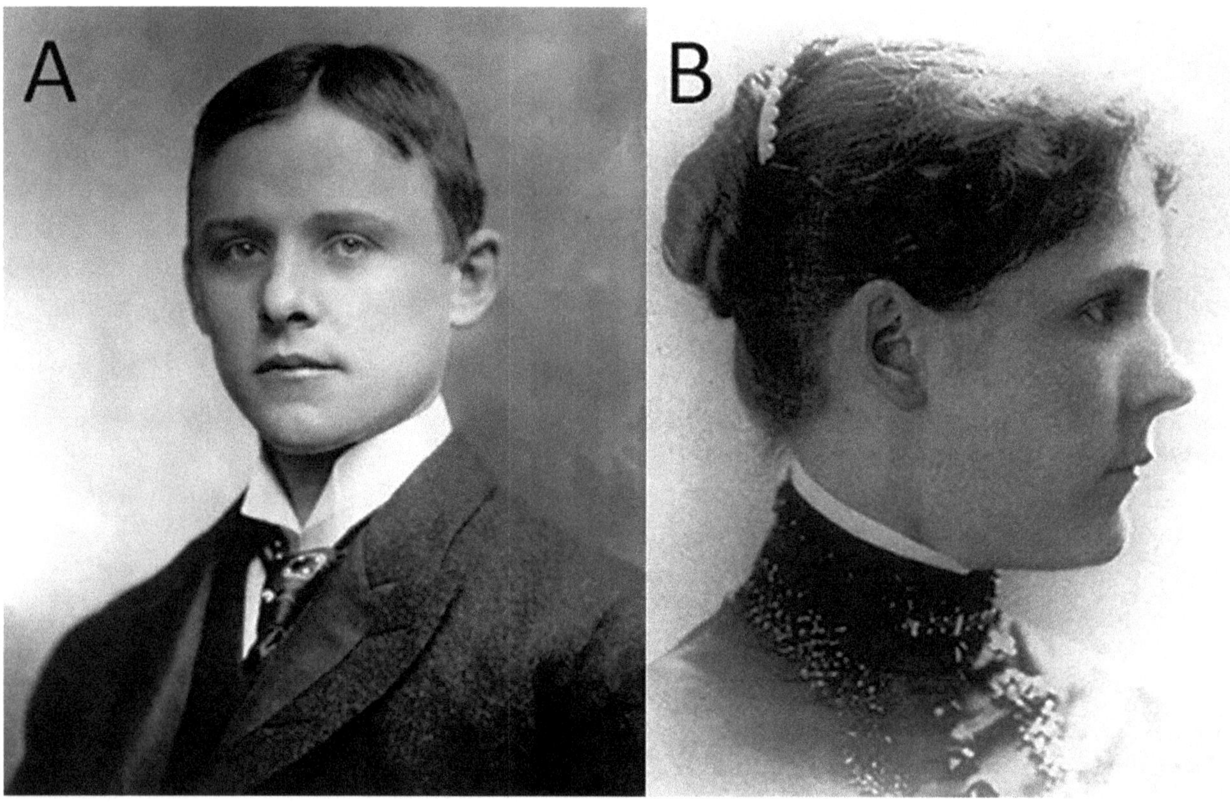

26.1. (A) Charles Martin Hall. (B) Julia Brainerd Hall, Charles's sister

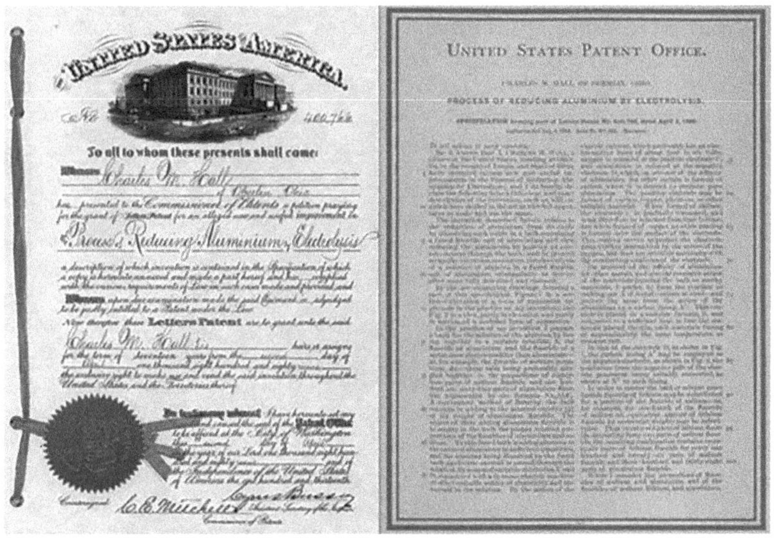

26.2. Charles Martin Hall's Aluminum patent (issued April 3, 1889)

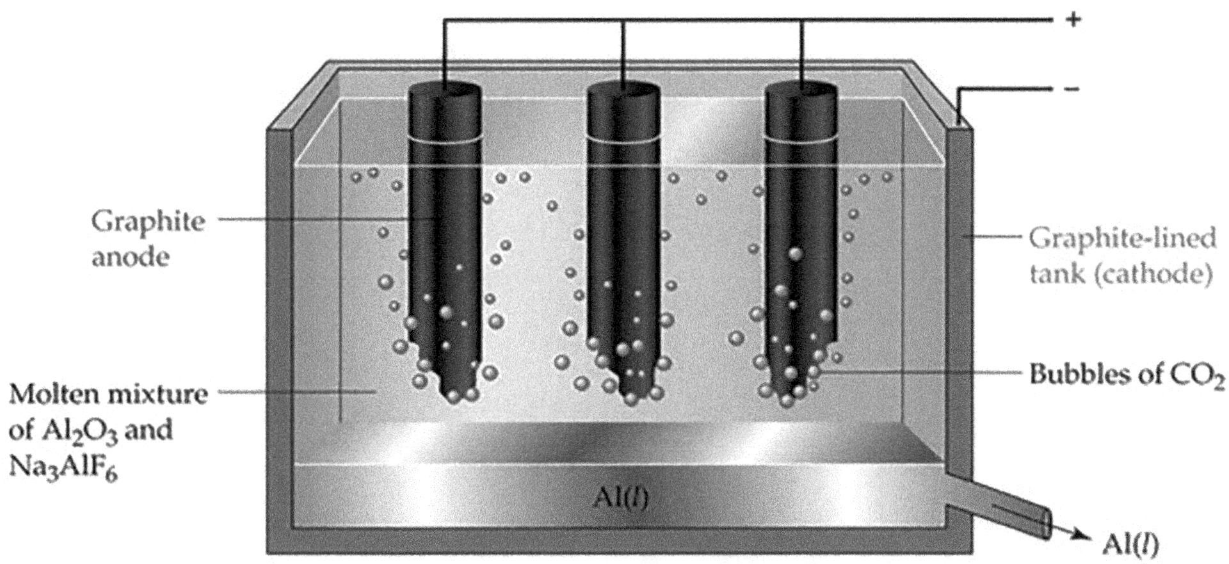

26.3. The scheme shows an electrolytic cell for production of aluminum by the Hall-Héroult process. Molten aluminum metal forms at the graphite cathode that lines the cell. Because molten aluminum is more dense than the Al_2O_3-Na_3AlF_6 mixture, it collects at the bottom of the cell and is drawn off periodically.

26.4. Drawing of Hall's original electrolytic cells

26.5. Aluminum electrolytic production at industrial scale (present time—aluminum plant in Massena, New York). Many electrolysers are shown in this photo.

27

PAUL LOUIS TOUSSAINT HÉROULT

Born: April 10, 1863, Thoury-Harcourt, France
Died: May 9, 1914, near Antibes, on the sea

Paul Louis Toussaint Héroult (27.1) was a French scientist who invented a commercial electrochemical production of aluminum, which was independently coinvented by him with Charles Martin Hall (American inventor, businessman, and chemist, 1863–1914) in 1886. Since the electrochemical production of aluminum was almost the same in the process developed by Héroult (in France) and Hall (in the USA), it was named Hall-Héroult process to acknowledge the contribution of both inventors.

In order to decrease the melting temperature of the aluminum oxide (Al_2O_3) and to make commercially profitable electrochemical production of metallic aluminum (Al) in a melted state (excluding water in the electrolyte), Héroult added cryolite (Na_3AlF_6). The same invention (except minor technical details) was made by Hall in the USA. Two very similar patents were obtained by Héroult in France (27.2) and by Hall in the USA in 1886.

The first aluminum plant based on Hall-Héroult technology in Europe was built in Neuhausen, Switzerland, on the banks of the Rhine. In 1888, the Societe Electrometallurgique Francaise was founded for the electrochemical production of aluminum, where Héroult became the technical director. Later, this company merged with the Societe Francaise de l'Aluminium Pur at which Héroult was also involved in research.

Héroult's second most important invention was the first commercially successful electric arc furnace (27.3) for steel production in 1900. The Héroult arc furnace replaced the giant smelters for the production of a variety of steels.

27.1. Paul Louis Toussaint Héroult

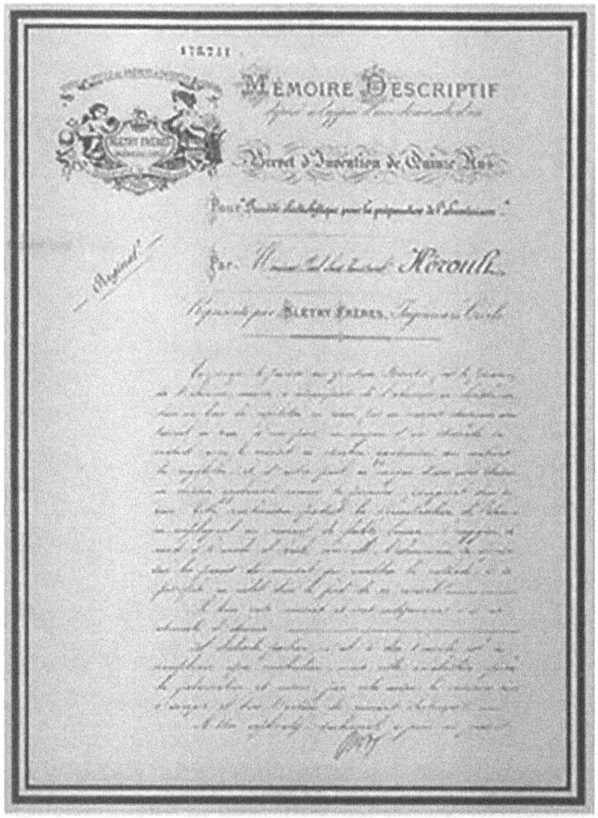

MÉMOIRE DESCRIPTIF

Brevet d'Invention de Quinze Ans

Par ... Héroult

Blétry Frères

27.2. Héroult's patent for the electrochemical production of aluminum obtained in France in 1886

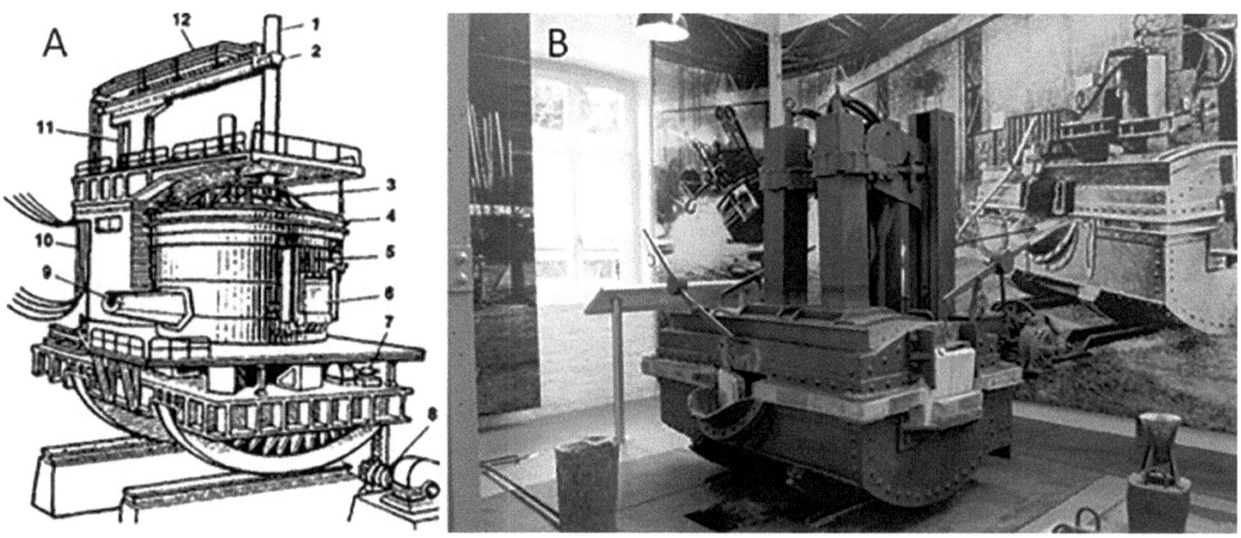

27.3. 1905 Héroult's electric arc DC furnace: (A) technical drawing and (B) photo of the electric arc in Deutschen Werkzeugmuseum, Remscheid, Germany

28

HANS LUGGIN

Born: June 24,1863, Klagenfurt, Austria
Died: December 5, 1899, Klagenfurt, Austria

Hans Luggin (28.1) was an Austrian physicist and chemist. He is well-known for the so-called Luggin capillary.

While doing electrochemistry experiments, Luggin suggested to use a special capillary (then named Luggin capillary) for connecting a reference electrode to the electrolyte solution close to the working electrode surface. The Luggin capillary served as a salt bridge between the reference and working electrodes. This bridge decreased the ohmic resistance and the corresponding potential drop (IR) between the working and reference electrodes. This allowed more precise potential measurements during electrochemical experiments.

Luggin capillary (28.2A) is usually a bent glass tube with a large enough opening to accommodate a reference electrode and a much smaller opening only large enough to ensure ionic conductivity. By locating the smaller opening close to the surface of a working electrode (28.2B), the influence of the IR drop can be minimized. However, it must not be placed too close to the working electrode; otherwise, the diffusion of redox species toward the working electrode might be blocked.

Another use of the Luggin capillary is its application in nonaqueous electrolyte solutions. Indeed, when the electrolyte solution is based on an organic solvent (e.g., dimethylformamide, DMF), the same nonaqueous solution can be applied in the Luggin capillary, separating the aqueous solution inside the reference electrode (e.g., calomel or Ag/AgCl reference) from the main nonaqueous electrolyte solution in an electrochemical cell. Importantly, this prevents water leakage from the aqueous reference electrode to the nonaqueous electrolyte solution in the cell (note that the nonaqueous solution may need to be dry).

Luggin also published on photoelectric effects, photography, electrocapillary effects, and on polarization of very thin metal membranes. However, his name is mostly associated with the capillary invented by him, even it was only a small technical improvement in electrochemical cells.

28.1. Hans Luggin

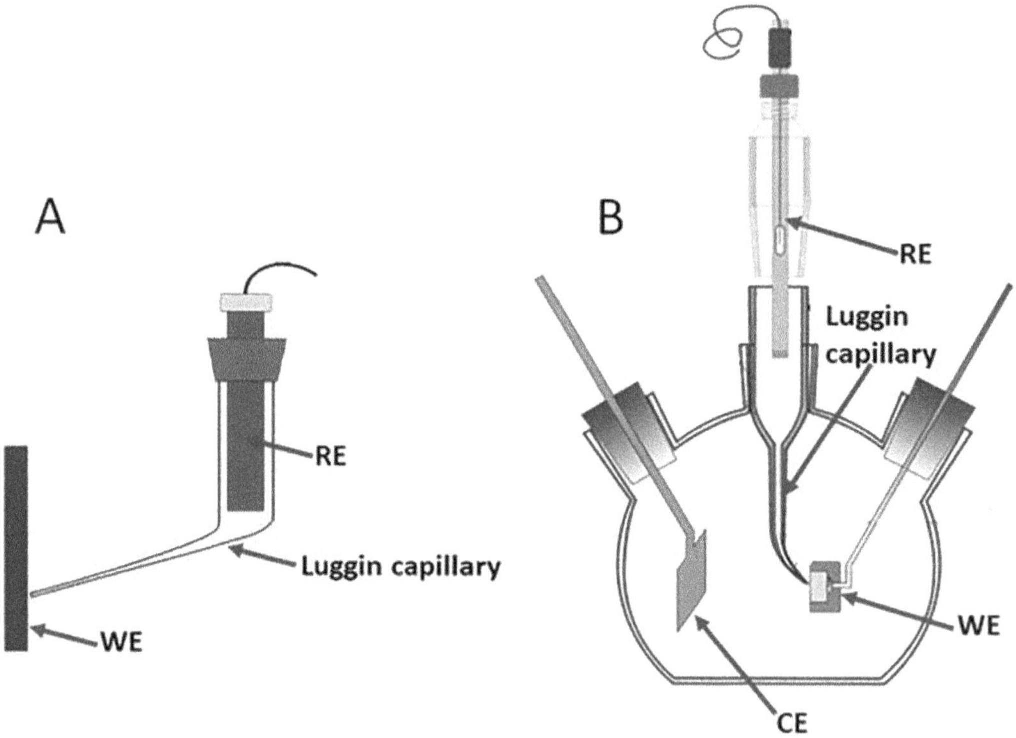

28.2. (A) Luggin capillary connecting a reference electrode to a working electrode. (B) An electrochemical cell with a Luggin capillary: WE (working electrode,) CE (counter electrode), RE (reference electrode)

29

HERBERT HENRY DOW

Born: February 26, 1866, Belleville, Ontario, Canada
Died: October 15, 1930, Rochester, Minnesota, USA

Herbert Henry Dow (29.1) was an American chemical industrialist who founded the American multinational conglomerate Dow Chemical. He was a prolific inventor of chemical processes and a successful businessman.

Dow invented industrial methods for production of various chemicals, notably the first electrochemical bromine (Br_2) extraction from seawater. Bromine, at that time, was a primary ingredient in medicines and was widely used in the newly established photographic industry (being used in the form of bromide salts). In 1890, he established Midland Chemical Company to produce bromine (29.2). The Dow process was the first commercial process for the production of bromine by electrolysis (patented in 1891). He also applied electrolysis to produce sodium hydroxide (NaOH) and chlorine gas (Cl_2) from a sodium chloride (NaCl) aqueous solution (29.3, 29.4). In 1895, he established the Dow Chemical Company, which became the fifth largest chemical company in the world producing bromine, chlorine, sulfur chloride, monochlorobenzene, phenol, magnesium, synthetic indigo among more than two thousand chemical products around the globe.

29.1. Herbert Dow (1888)

29.2. The photo shows Evens Mill, rented by Herbert Dow in 1890 as the location of his new enterprise, the Midland Chemical Company. Dow chose the grist mill because it provided his basic requirements: a source of brine and a source of electrical energy. Here, Dow did his early experiments in bromine and chlorine production from brines.

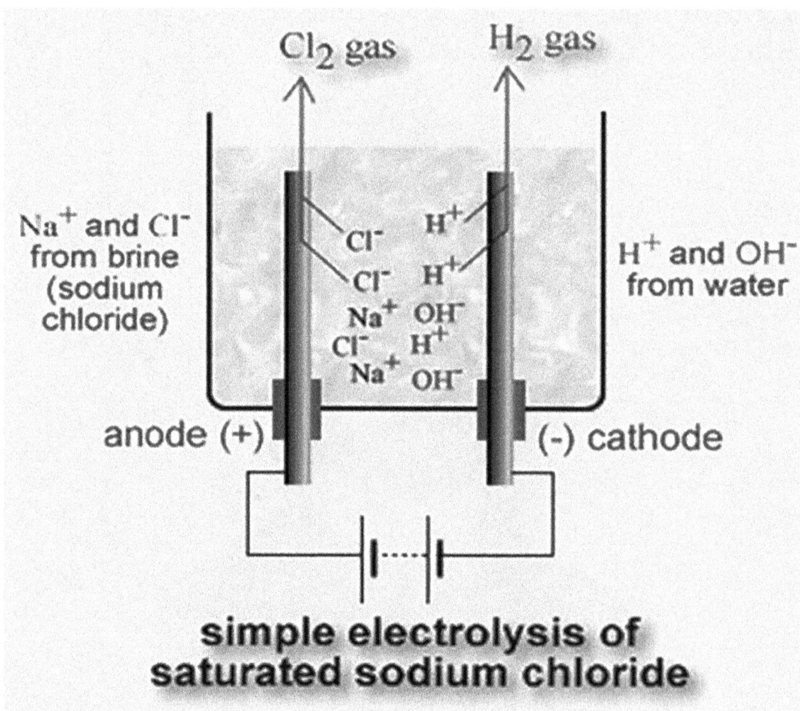

29.3. Schematics of the NaCl aqueous solution electrolysis to yield NaOH and Cl_2

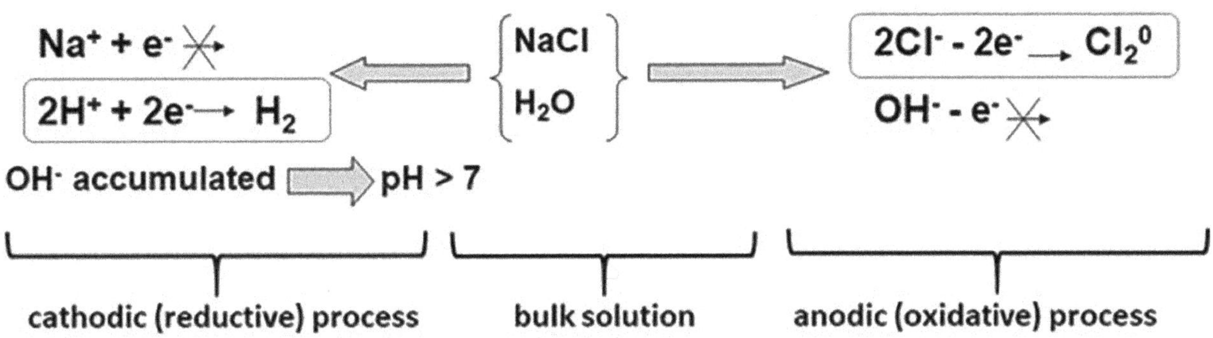

29.4. Chemical reactions proceeding during the electrolysis of a NaCl aqueous solution

30

FREDERICK GEORGE DONNAN

Born: September 5, 1870, Colombo, Ceylon, British Empire (presently Sri Lanka)
Died: December 16, 1956, Canterbury, Kent, United Kingdom

Frederick George Donnan (30.1) was a British-Irish physical chemist who is known for his work on equilibria at a semipermeable (particularly biological) membrane (known as Donnan equilibrium). His work was particularly important for describing ionic transport in biological cells (30.2).

Donnan greatly contributed to the development of colloid chemistry, physical chemistry, and electrochemistry. He studied (1911) the electrical potential setup at a semipermeable membrane between two electrolytes, an effect of great importance in biological cells. This equilibrium results in the formation of Donnan potential across a membrane.

The equilibrium potential can be established on a semipermeable membrane through which the solvent and some, but not all, of the dissolved ions can pass (30.2). The membrane prevents some ion species (e.g., macromolecules such as proteins or other large biomolecules) to transfer through and results in changing the composition of small molecular ions in solutions on both sides of the membrane, producing a potential difference across the membrane. The formation of the ionic equilibrium on both sides of the membrane and the generation of the electric potential difference across the membrane are both associated with Donnan's name.

30.1. Frederick George Donnan

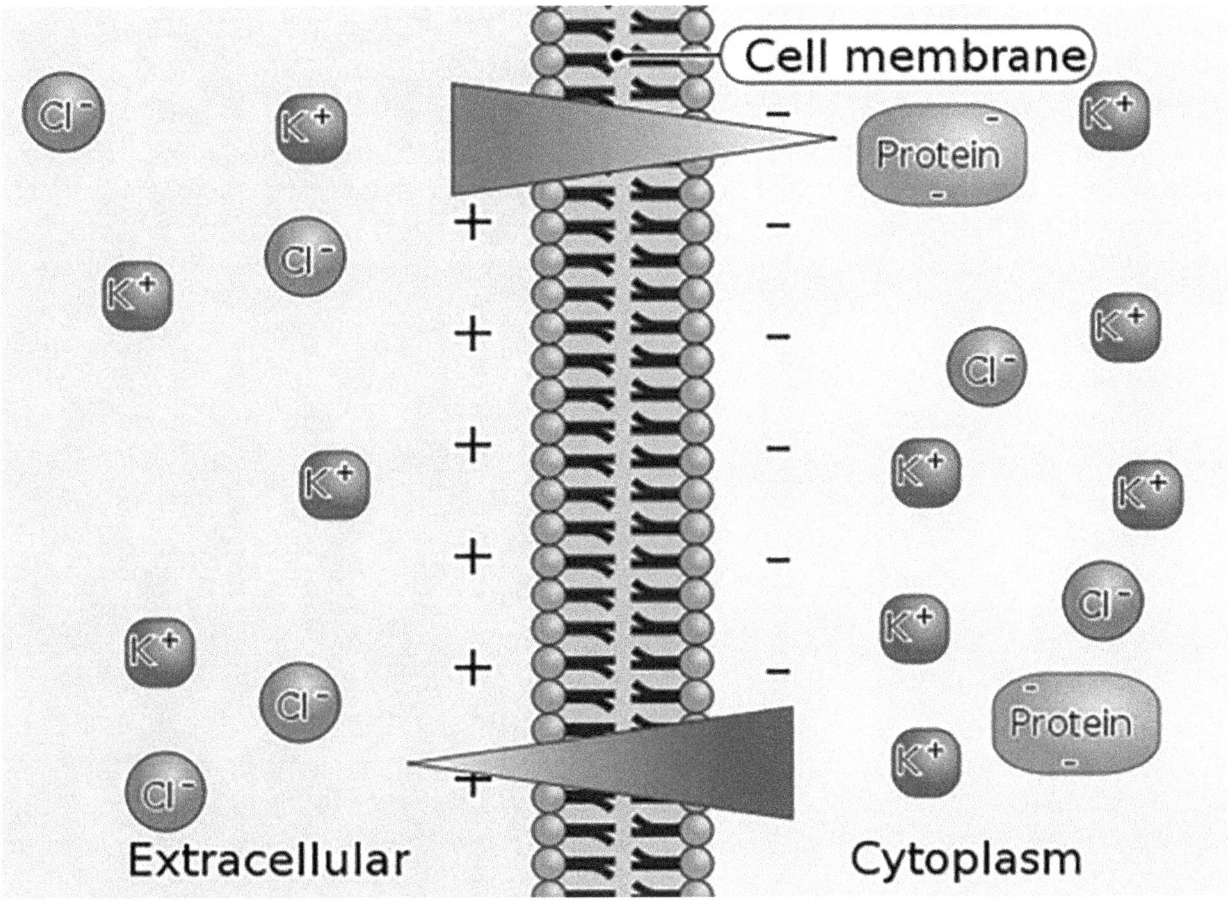

30.2. Schematically shown Donnan equilibrium formation across a cell membrane. The presence of a charged impermeable macro-size ion (for example, a protein) on one side of a membrane will result in an asymmetric distribution of permeable charged ions (represented in this scheme as K^+ and Cl^-), then generating a potential difference across the membrane. The triangles show schematically the concentration gradients for Cl^- and K^+.

31

FREDERICK GARDNER COTTRELL

Born: January 10, 1877, Oakland, California, USA
Died: November 16, 1948, Berkeley, California, USA

Frederick Gardner Cottrell (31.1) was an American physical chemist, inventor, and philanthropist. He is best known for his invention of the electrostatic precipitator, one of the first inventions designed to eliminate air pollution, and for the famous Cottrell equation describing time-dependent current during diffusion-controlled chronoamperometric experiments.

In 1907, Cottrell applied for a patent on a device for charging air-suspended particles and collecting them through electrostatic attraction. It was the first electrostatic precipitator (31.2). The first application of the precipitator was a collection of sulfuric acid mist and lead oxide fumes emitted from various acid-making and smelting processes.

Important for electrochemistry, the Cottrell equation (derived in 1903) describes electric current changes during a control-potential chronoamperometry experiment (31.3) performed with a diffusion-controlled redox process of soluble species. The equation shows reciprocal square root dependence, $I \propto \sqrt{1/t}$, of the current as a function of time after a potential step.

$$I = n F A c_0 \sqrt{\frac{D}{\pi t}}$$

I = current, in units of A (ampere); n = number of electrons participating in the redox process per molecule; A = area of the (planar) electrode in cm^2; c_0 = bulk concentration of the redox species; F = Faraday constant, 96485 C/mol; D = diffusion coefficient for the redox species in cm^2/s, t = time in *s*.

For a simple redox reaction, not complicated by reactant adsorption or by secondary reactions (e.g., ferrocene/ferrocenium redox reaction), the current measured depends on the rate at which the analyte diffuses to the electrode—that is, the current is said to be diffusion-controlled. The classical version of the Cottrell equation corresponds to the diffusion of reactants to a planar electrode, which is reciprocal square root dependent on time. Other kinds of diffusion to spherical, cylindrical, and rectangular electrode geometries result in different versions of the Cottrell equation, which can be derived using the corresponding Laplace operator and boundary conditions in conjunction with Fick's second law of diffusion. Notably, if the redox species are immobilized at an electrode surface (in the case of chemically modified electrodes) or if the electrochemical process is complicated with the adsorption of the redox species, the Cottrell equation is not applicable because the process is not diffusion-controlled.

31.1. Frederick Gardner Cottrell (ca. 1910)

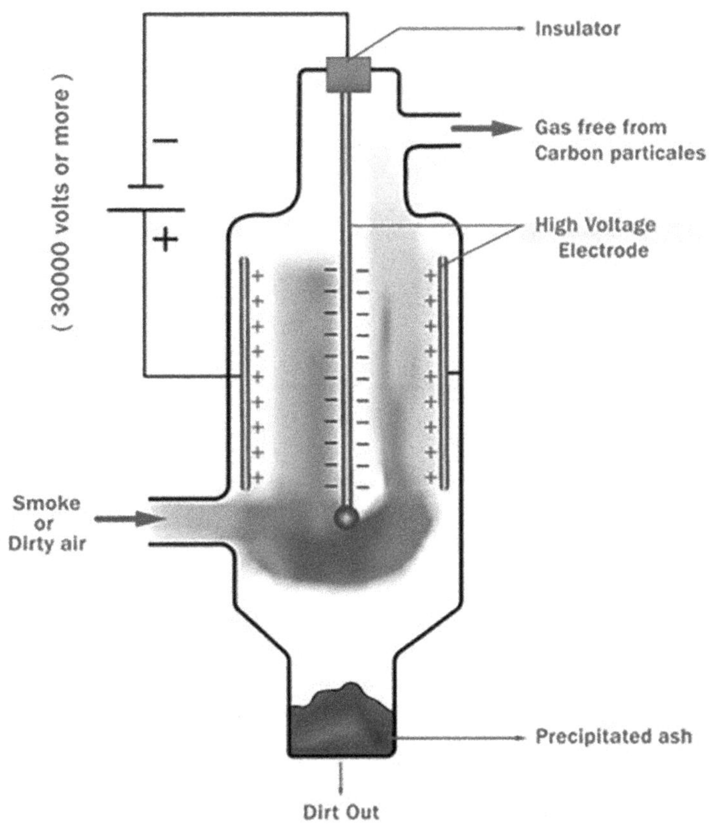

31.2. Cottrell's electrostatic precipitator shown schematically

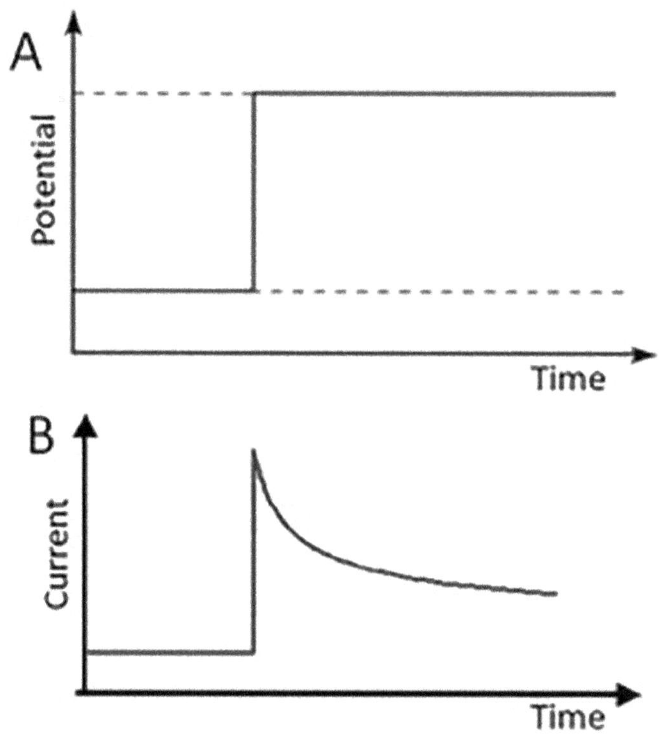

31.3. Chronoamperometry experiment: (A) potential step applied on a working electrode and (B) current response to the potential step

32

CARL TUBANDT

Born: December 3, 1878, Halle Germany
Died: January 17, 1942, Berlin, Germany

Carl Tubandt (32.1) was a German scientist, a pioneer of solid state electrochemistry. Tubandt introduced a methodology for determining the transport numbers of ions in solid electrolytes, which is known as Tubandt method. This method allowed the finding of conductivities in solid electrolytes, and the mechanism of conductivity was explained. The transport numbers indicated the ionic species dominating the electrical conductivity in solids. The research was particularly performed with solid electrolytes such as silver, lead, and copper halides, and silver sulfide. The experiments with solid silver bromide (AgBr) performed in the dark (to exclude the photoelectric effect, resulting in the increase in conductivity) demonstrated that silver ions (Ag^+) are mostly responsible for the ionic conductivity in this material. Surprisingly, it was found that this solid material has higher conductivity than its melt.

The Tubandt method was based on pressing the solid electrolyte in tablets and passing current through the pile of tablets (32.2). Weighing the tablets following the passage of a certain charge allowed the calculation of the transport numbers. The Tubandt method is equally applicable to single-crystal and polycrystalline materials. Problems may only be caused by high resistance between the tablets. Scientifically, the Tubandt method is based on Faraday's low connecting the electric charge passed and the mass change upon electrochemical reactions.

32.1. Carl Tubandt

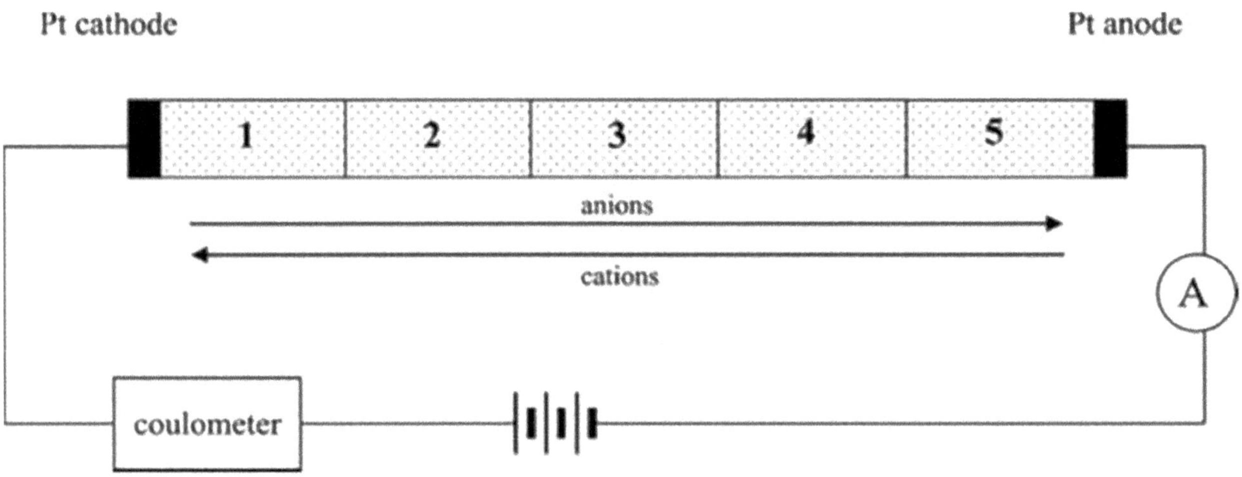

32.2. The scheme illustrating the Tubandt method for analysis of conductivity and finding transport numbers in solid materials. Numbers 1–5 in the scheme correspond to the tablets made of solids with ionic composition (e.g., AgCl).

33

GYÖRGY HEVESY

Born: August 1, 1885, Budapest, Austro-Hungarian Empire (presently Hungary)
Died: July 5, 1966, Freiburg im Breisgau, Germany

György Hevesy (33.1) was a Hungarian (with Jewish ethnicity) chemist/radiochemist. Hevesy performed his research in different European countries, being forced to move by the German Nazi regime.

Hevesy performed research mostly with radioactive isotopes and received Nobel Prize in Chemistry in 1943 for his key role in the development of radioactive tracers to study chemical and biochemical processes. He developed the X-ray fluorescence analytical method and codiscovered with Dirk Coster (Dutch physicist, 1889–1950) element hafnium ($_{72}Hf$). His important contribution to electrochemistry was the use of the radiotracer technique for the analysis of mechanisms of electrochemical processes. The application of radiotracer methods in electrochemistry was initiated by Hevesy in 1914. The aim of these studies was to demonstrate that isotopic elements can replace each other in both electrodeposition and equilibrium processes.

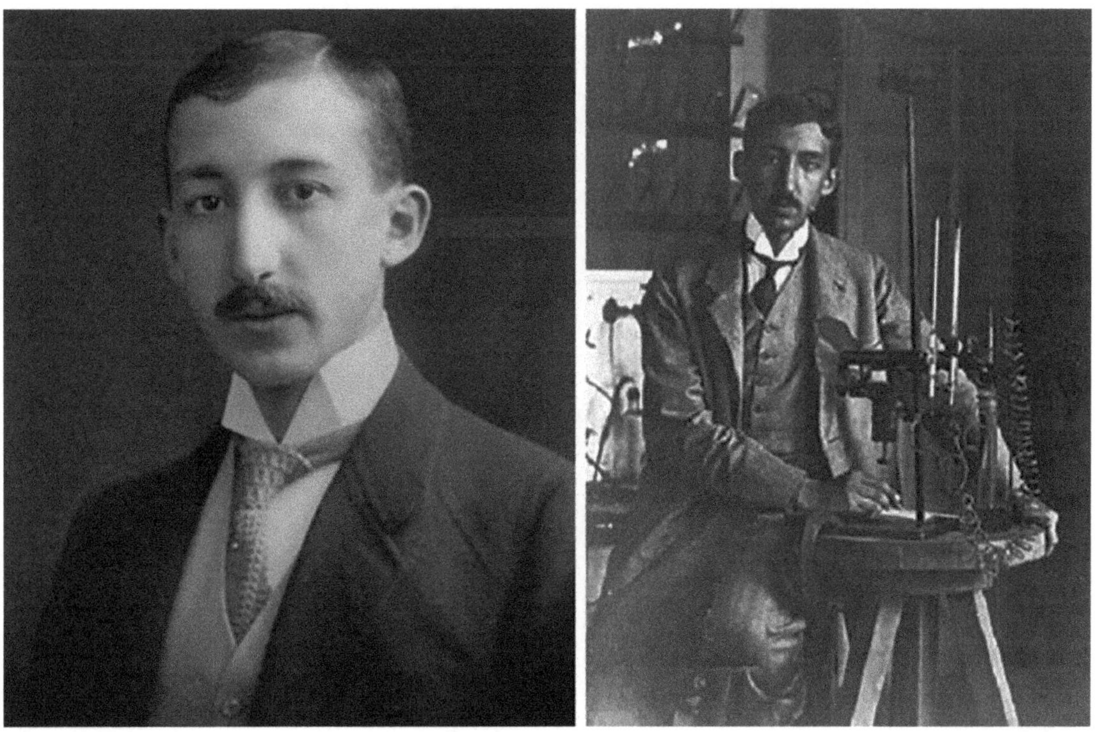

33.1. György Hevesy

34

JAMES CLOYD DOWNS

Born: November 6, 1885, Eugene, Oregon, USA
Died: December 11, 1957, USA

James Cloyd Downs (34.1) was an American chemical engineer who is known for the invention of the Downs cell for electrolytic production of metallic sodium (Na).

In 1923, Downs invented (patented in 1924) an efficient electrochemical method for the commercial production of metallic sodium. The electrolysis was performed in a molten NaCl (with $CaCl_2$ addition for decreasing the melting temperature) in a special apparatus called Downs cell (34.2). The calcium ions (Ca^{2+}) are not reduced in the electrolysis process because their reduction potential is lower than that of sodium. The cell design keeps the produced metallic sodium (Na) and chlorine (Cl_2) apart so that they cannot react with each other. The Downs's process also produces chlorine as a by-product, although chlorine produced this way accounts for only a small fraction of chlorine produced industrially by other methods (mostly by electrolysis of NaCl aqueous solution).

34.1. James Cloyd Downs with wife, Mabel, and grandson, Tom Downs (2009)

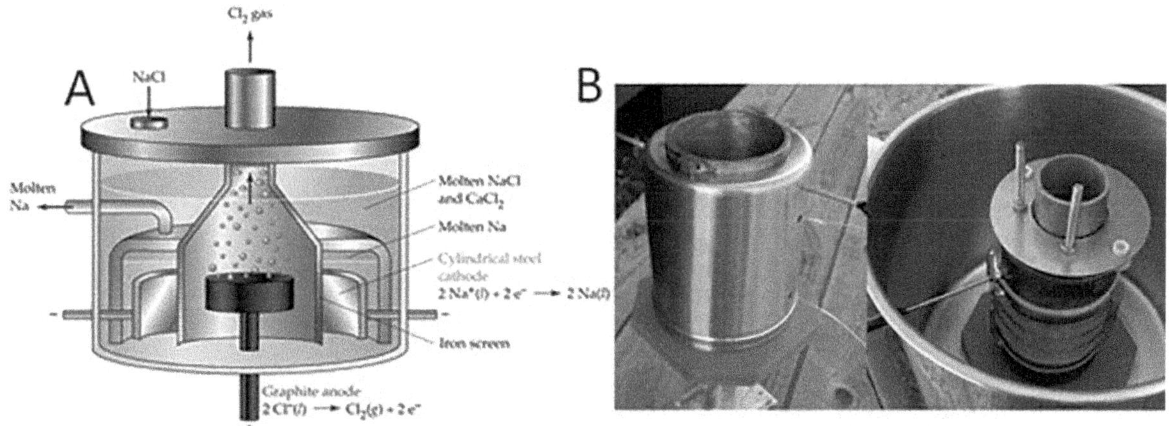

34.2. (A) Schematics of the Downs cell operation and (B) the photo of the open Downs cell

35

MAX VOLMER

Born: May 3, 1885, Hilden, Germany
Died: June 3, 1965, Potsdam, Germany

Max Volmer (35.1) was a German physicist remembered for his important contributions to the physical chemistry of nucleation and growth of new phases, especially crystals. His another important work was in the area of theoretical electrochemistry, particularly in formulating famous Butler-Volmer equation connecting current and voltage in electrochemical systems.

Volmer, together with his student Erdey-Gruz, published a paper on the theory of hydrogen overpotential, which resulted in the Butler-Volmer equation:

$$i = i_0 \cdot \left[\exp\left(\frac{\alpha_A nF}{RT}\eta\right) - \exp\left(-\frac{\alpha_C nF}{RT}\eta\right)\right]$$

This equation is named, also acknowledging John Alfred Valentine Butler (British physical chemist, 1899–1977), who independently invented a similar mathematical expression connecting currents and voltages applied. The Butler-Volmer equation was first published by John Alfred Valentine Butler in 1919 and later extended by Max Volmer in 1923. The Butler-Volmer equation is derived from the laws of thermodynamics and describes the relationship between the current passing through an electrode and the potential difference across it. The equation takes into account the activation overpotential, which is the energy required for the charge transfer reaction to occur at the electrode surface. The Butler-Volmer equation is a fundamental equation in electrochemistry that plays a crucial role in modeling and understanding electrode processes.

35.1. Max Volmer

36

ZYGMUNT ALEKSANDER KLEMENSIEWICZ

Born: June 24, 1886, Krakau, Austro-Hungarian Empire (now Krakow, Poland)
Died: March 25, 1963, Gliwice, Poland

Zygmunt Aleksander Klemensiewicz (36.1) was a Polish physicist and physical chemist who made a pioneering contribution to the development of a glass electrode.

Many scientists devoted their whole lives to develop specific areas of science or technology until their hard work and commitment resulted in success. There are only a few examples when success was achieved rapidly and particularly at a young age. The success story of Zygmunt Aleksander Klemensiewicz is a very rare example of such luck.

Zygmunt was a young Polish postdoctoral researcher when he arrived in 1908 to the group of Professor Fritz Jacobus Haber (German scientist, 1868–1934) at Technische Hochschule Fridericana (Technical University) in Karlsruhe, Germany. He got his first assignment to investigate the electrochemical behavior of glass in acidic and basic solutions. By that time, some preliminary data already existed but without consistent information and with no straightforward dependence between the solution composition (particularly the pH vale) and electrochemical response of a glass material. Because of the high resistance of the used glass, the response was measured as a potential variation while the current was obviously not measurable. The results were strongly dependent on the glass composition, its pretreatment, and thickness.

Klemensiewicz made an electrode (36.2) consisting of a glass pipe 7–8 mm in diameter and 8 cm long, ending with a glass bulb of 2.5 cm in diameter and 0.06–0.1 mm thick. The sample that he used was easily meltable glass from Thüringen. Immediately after blowing out the glass bulb, it was treated with water vapor from both sides and then soaked in distilled water (moisturizing) before use or between the potential measurements. This preparation procedure was critically important for obtaining good results, and it was found experimentally by chance. In other words, Klemensiewicz was lucky enough to find this procedure rather soon without numerous attempts with many other procedures. Importantly, the results were particularly well reproducible when the whole titration curve with varied pH was measured. The titration curves demonstrated around 0.5 V potential difference upon pH variation when a 0.1 M NaOH solution was titrated with a 0.1 M HCl solution. Single measurements for one specific pH value were much less reproducible and demonstrated large fluctuations for the measured potentials.

Professor Haber hardly believed that his junior assistant obtained so good results in a such short time. Soon after the experiments, on January 28, 1909, the results were reported at the meeting of the Chemical Society in Karlsruhe. Then a scientific paper composed of two parts (theoretical and experimental) was published in a German journal titled *Zeitschrift für Physikalische Chemie*. The theoretical part was written and signed by Haber. (Notably, Klemensiewicz did not fully agree with the theoretical part; thus, he didn't sign it.) The experimental part was signed by both Klemensiewicz and Haber. For Klemensiewicz, it was one of the most important publications in his life. For Haber, it was also important although he had already written a lot of papers before 1909, not mentioning a later work on ammonia synthesis for which he got a Nobel Prize in Chemistry in 1918. It is interesting to mention that this was the only big success in the whole professional life of Klemensiewicz.

Overall, the unusually fast result obtained by Klemensiewicz in the lab of Haber became an important background for the future work on glass electrodes by other scientists. Finally, all pH-sensitive and ion-selective glass electrodes have been based on the results obtained by Klemensiewicz and Haber.

36.1. Zygmunt Aleksander Klemensiewicz (before 1938). Note that the important work on the glass electrodes was performed by Klemensiewicz when he was much younger than shown in this photo.

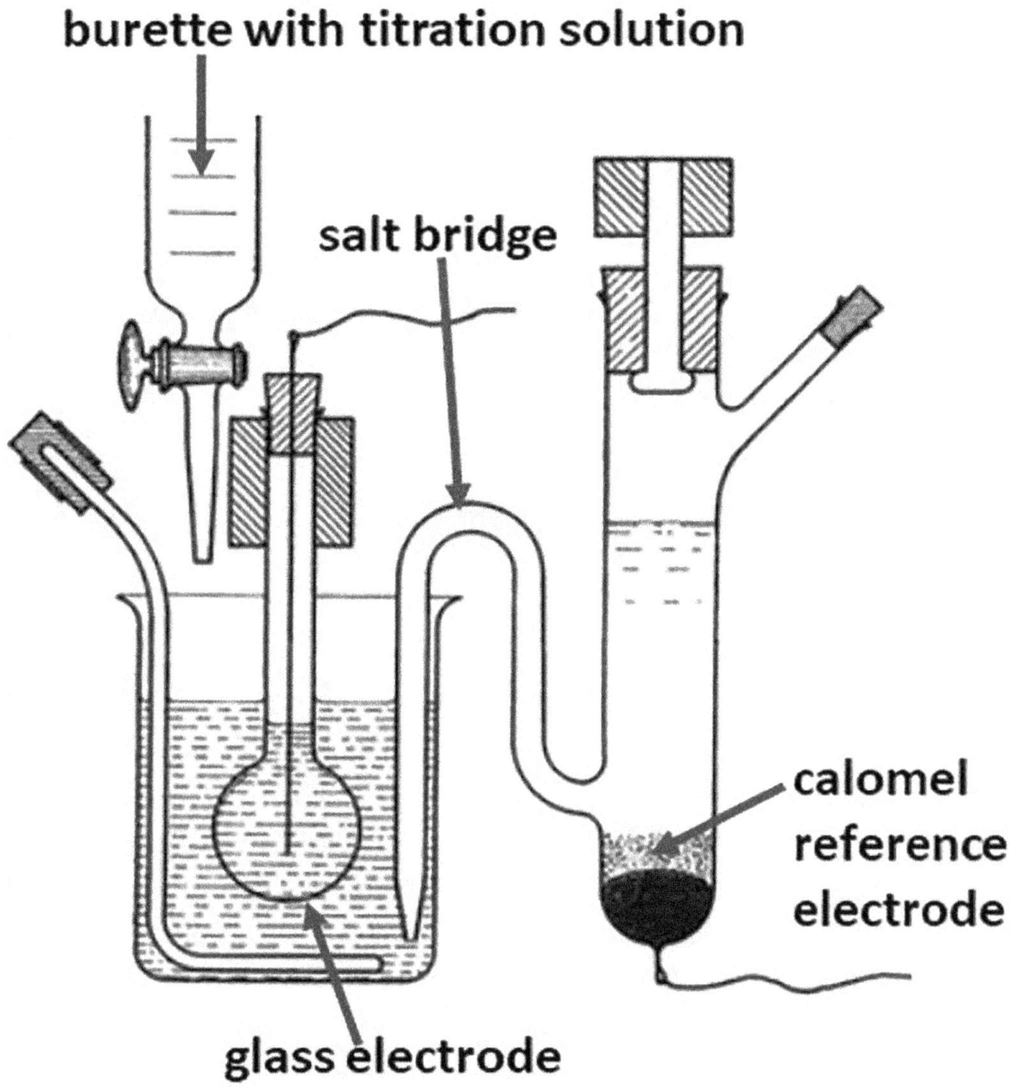

36.2. The electrochemical cell used by Klemensiewicz for measuring potentials of a glass electrode upon the base-acid titration

37

JAROSLAV HEYROVSKÝ

Born: December 20, 1890, Prague, Austro-Hungarian Empire
Died: March 27, 1967, Prague, Czechoslovakia (now Czech Republic)

Jaroslav Heyrovský (37.1) was a Czech scientist and a pioneer of electroanalytical chemistry. He invented polarography—the very first electroanalytical chemistry method, precursor of all modern electroanalytical methods.

One of the first scientific projects performed by Heyrovský was the study of surface tension at the interface of a dropping mercury electrode as the function of the potential applied. The electrode was composed of a capillary filled with liquid mercury freely dropping from the capillary open end. The manually varied voltage was applied between a mercury electrode and a metallic counter electrode. This was a very tedious experiment. The weight of an individual mercury droplet was measured for each electric voltage applied. The first simplification in the experimental procedure made by Heyrovský was the measurement of the droplet lifetime (in other words, dropping period) instead of the droplet's weight.

The dropping period was dependent on the droplet surface tension, which in turn was a function of the electric potential applied. There was only one step left to the discovery of polarography, and this step was made by Heyrovský when he decided to measure the electric current produced on the electrode instead of the dropping time. Indeed, the current was a function of the mercury droplet size, which was also a function of the potential applied. Still, Heyrovský experienced two technical problems: the lack of a sensitive enough instrument for measuring the current (galvanometer that time) and manual variation of the voltage applied (using rheostat—a variable resistor).

Rapid progress in electrical instrumentation helped Heyrovský to find the sensitive galvanometer, and the most important technical improvements were made by Masuzo Shikata (Japanese chemist, 1895–1964, a research assistant in the Heyrovský group, similar to a postdoc now) who designed in 1924 an instrument for automatic increase of the voltage applied on the electrodes (37.2). The important scientific outcome was obtained when redox species were added to the electrolyte solution. Their redox reactions (reduction or oxidation, depending on their redox potentials and voltage applied) increased the measured current. This was the beginning of polarography, the first electroanalytical method (37.3).

At a voltage specific for different substances, redox reactions cause the current to rapidly increase to a level dependent on the concentration of the substance (37.4). The measured current vs. voltage function allowed deriving a lot of information about the redox species: their redox potential, number of electrons involved in the redox process, reversibility of the redox transformation, etc. Although Heyrovský and Shikata did not patent the instrument, they obtained a copyright on the name they coined for the technique and the instrument—*polarography* and *polarograph*, respectively.

In 1959, Heyrovský received the Nobel Prize in Chemistry for his invention and development of the polarographic method of analysis. Jaroslav Heyrovský devoted all his life to the polarography study and organized a large group of scientists working on polarography, particularly in Prague, where the J. Heyrovsky Institute of Physical Chemistry keeps his name in the institute title.

37.1. Jaroslav Heyrovský with a polarographic stand

37.2. (A) The original polarograph of J. Heyrovský and M. Shikata (1924). (B) This instrument is in the exhibition in the museum of J. Heyrovský Institute of Physical Chemistry.

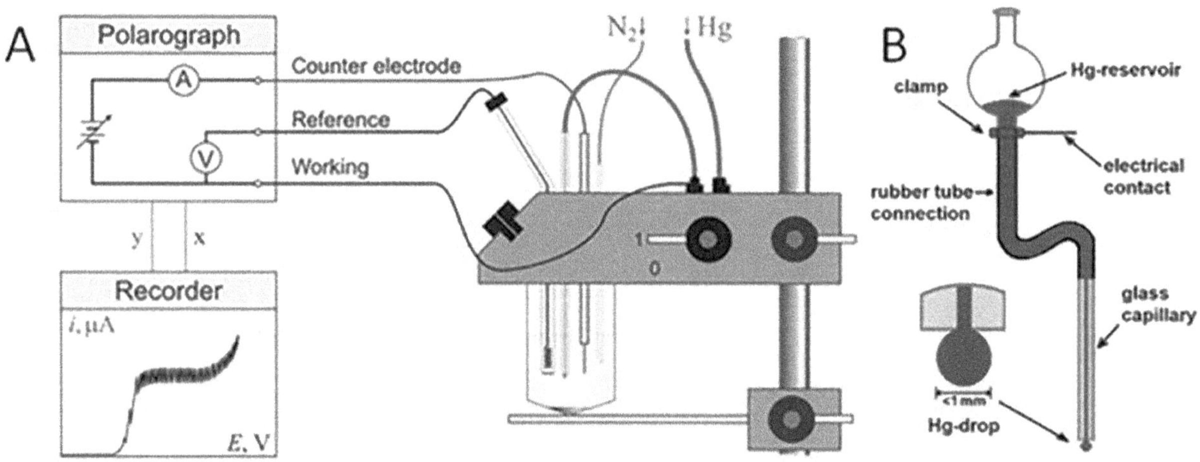

37.3. (A) Schematics of the polarographic measurements invented by Heyrovský. (Note that the use of a reference electrode became possible when potentiostates were invented; the original Heyrovský's measurements were performed with two electrodes—working and counter, without a reference electrode). (B) Dropping mercury electrode.

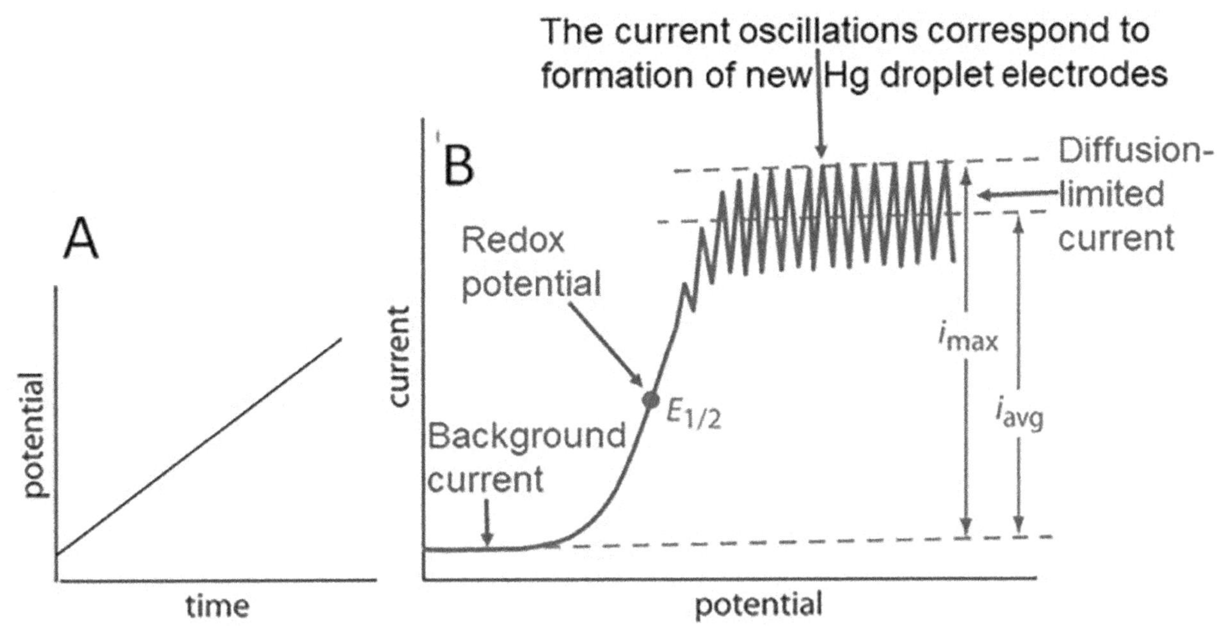

37.4. Typical polarogram. (A) Potential applied on a dropping Hg electrode is increasing linearly with time. (B) Current measured on a dropping Hg electrode is changing, reflecting redox transformations of soluble species at the Hg electrode surface.

38

IZAAK MAURITS (PIET) KOLTHOFF

Born: February 11, 1894, Almelo, Netherlands
Died: March 4, 1993, St. Paul, Minnesota, USA

Izaak Maurits (Piet) Kolthoff (38.1) was a Dutch-born American scientist who particularly contributed to analytical chemistry, including electroanalytical chemistry. He was also a great chemistry educator. He is widely considered as the father of analytical chemistry for the large volume of published research, particularly textbooks (38.2), in diverse fields of analysis.

Professor Kolthoff's research was in many areas of chemistry, particularly focused on analytical chemistry. His work provided a solid scientific foundation for modern analytical chemistry. The research performed by Kolthoff transformed analytical chemistry from a random collection of empirical recipes and prescriptions to a fundamental branch of modern chemistry. Kolthoff studied many different analytical methods—including acid-base analysis, gravimetric analysis, volumetric analysis, iodometry, the theory of colloids, and crystal growth—thereby establishing a scientific basis for gravimetry. His work with acids and bases in which he was the first to fully apply Arrhenius theory of electrolytes established the rational choice of end point indicators and transformed titrimetry from a practical art to an exacting science. He developed the theory of potentiometric and conductometric titrations.

While electrochemistry (particularly in a narrow definition) was not his primary interest, Kolthoff performed important work in polarography–electroanalytical method with the use of a dropping mercury electrode (DME). His work on polarography was inspired by visiting Jaroslav Heyrovský (Czech electrochemist and inventor of polarography) in 1933. Later, two graduate students of Kolthoff became important scientists contributed to the development of electroanalytical chemistry: James J. Lingane (American electrochemist, 1909–1994) worked on polarography using DME, and Herbert "Herb" August Laitinen (American electrochemist, 1915–1991) worked with solid micro-electrodes. Their studies were extremely important for the development of electroanalytical chemistry, and the results were summarized in seminal review papers and books. It is important to mention that Lingane and Laitinen constructed advanced electroanalytical instrumentation.

38.1. Izaak Maurits (Piet) Kolthoff, 1952

38.2. Examples of Kolthoff's books

39

ALEXANDER NAUMOVICH FRUMKIN

Born: October 24, 1895, Kishinev, Russian Empire (now Chișinău Moldova)
Died: May 27, 1976, Moscow, USSR (now Russia)

Alexander Naumovich Frumkin (39.1) was a Soviet/Russian (with Jewish ethnicity) electrochemist. He was the most famous and most important scientist in the area of electrochemistry. Frumkin was the father of Russian electrochemistry. He created a novel school of electrochemistry and organized the famous academic Institute of Electrochemistry in Moscow, later named after him the Frumkin Institute of Electrochemistry. (Note that the author of this book obtained PhD in this Institute in 1983.)

The most fundamental achievement of the outstanding work of Frumkin and his school (39.2 left) was the creation of the fundamental theory of electrode reactions, which considers the influence of the structure of the electrode/solution interface on the rate of the elementary electron transfer event—the theory of the slow discharge. Later, this theory received confirmation and development within the framework of the contemporary physical models of the electron transfer and reorganization of the polar medium. Frumkin introduced the concept of the zero charge potential, the most important characteristic of a metal surface. Volta's problem about the nature of the electromotive force (EMF) of electrochemical circuits, which was the subject of scientific discussion for more than 120 years, was successfully solved using the approach developed by Frumkin.

Adsorption at an electrode surface was considered by Frumkin, resulting in the formulation of Frumkin isotherm. Frumkin's isotherm that describes adsorption phenomena taking into account interactions between the adsorbed species became the well-known and important development of Langmuir's isotherm. To make a realistic model of adsorption, Frumkin assumed that there was some interaction between the adsorbate species, attractive or repulsive, resulting in the isotherm equation different from Langmuir's isotherm.

Novel experimental methods that became classical were developed under the supervision of Frumkin. Many applied electrochemical projects—for instance, projects related to chemical sources of electrical power, industrial electrolysis, protection against corrosion, and more—were successfully developed under the supervision of Frumkin. The most famous Frumkin's book *Kinetics of Electrode Processes* was published in 1952 (39.2, right).

Frumkin's work provided the foundation stones of modern electrochemistry. Particularly, he integrated the theoretical description of homogeneous catalysis and electrochemical kinetics. Then he considered jointly the fields of surface chemistry and electrode reactions. This allowed him to explain the salt effect on electrode kinetics. Overall, the whole area of physical, analytical, and applied electrochemistry is based on the theoretical and experimental work performed by Frumkin and his collaborators, particularly at the Frumkin Institute of Electrochemistry in Moscow.

Frumkin had very broad education and various interests. He loved painting, poetry, and history. He passionately loved journeys, particularly in the mountains (39.3). Academician Semenov and other outstanding scientists were Frumkin's companions on these journeys. Besides literature, Frumkin was also enthusiastic about pictures, and his colleagues celebrated his seventieth birthday in 1965 with pictures of him in the styles of Gauguin (by A. A. Lopatkin) and Picasso (by O. A. Petrii and R. V. Marvet) (39.4).

39.1. Alexander Naumovich Frumkin

39.2. Frumkin (*sitting at the right*) with the colleagues discussing scientific results (*left*). The most famous Frumkin's book, coauthored by his colleagues, "Kinetics of Electrode Processes," was published in Russian in 1952 (*right*).

39.3. A. N. Frumkin in Chegem mountains (1938)

A B

39.4. Alexander Naumovich Frumkin in art created by his friends and colleagues. (A) Portrait of Frumkin in Picasso's style. Graphically represented electrochemical functions studied by Frumkin are easily recognized by electrochemists. Some of these typical electrochemical plots have been collected in the portrait of Frumkin painted in Picasso's manner. In this portrait, you can also find two traditional electrochemical components: a capillary with a moving mercury drop that results in the polarographic maximum and a rotating ring-disk electrode. Frumkin has greatly contributed to the development of these techniques. (B) *Frumkin on Tahiti island* painted in the Gogen's style. Frumkin greatly contributed to the creation and development of scientific societies in many regions and cities of the USSR, particularly in Tartu, Riga, Alma Ata, Tbilisi. The theme of Frumkin-traveler inspired Lopatkin to paint Frumkin's portrait in the Gogen's manner.

40

HANS FALKENHAGEN

Born: May 13, 1895, Wernigerode, Germany
Died: June 26, 1971, Rostock, Germany

Hans Falkenhagen (40.1) was a German physicist and electrochemist best known for explanations of the Debye-Falkenhagen effect and Wien effect.

While working with Peter Joseph William Debye (Dutch-American physicist and physical chemist, 1884–1966), Falkenhagen developed a theory of electrolyte viscosity, particularly explaining Wien effect—an experimentally observed increase in the conductivity of electrolytes at very high electrical potential gradients. The effect was named after Max Karl Werner Wien (German physicist, 1866–1938).

The most important theoretical work performed by Falkenhagen and Debye explains the increase in the conductivity of an electrolyte solution when the applied voltage has a very high frequency. This is known as the Debye-Falkenhagen effect. The conductivity dependence on the voltage frequency was explained by the existence of an ionic atmosphere with a finite time of relaxation. The theoretical predictions made by Debye and Falkenhagen in 1929 were fully confirmed by experiments with impedance measurements.

40.1. Hans Falkenhagen

41

MASUZO SHIKATA

Born: August 10, 1895, Tokyo, Japan
Died: May 8, 1964, Kyoto, Japan

Masuzo Shikata (41.1) was an associate of Prof. Jaroslav Heyrovský (Czech electrochemist and inventor, 1890–1967) with whom he developed the first polarograph—the first automatic instrument to record current-voltage dependencies on a dropping mercury electrode. This polarograph was the precursor of all electrochemical analyzers used until now.

In 1921, Shikata received a scholarship from Japan's Ministry of Education, which allowed him to stay in Germany and study surface chemistry under Prof. Isidor Traube (German physical chemist, 1860–1943) at Technische Hochschule in Berlin. At this time, Heyrovský started original and fruitful research applying a dropping mercury electrode for electrochemistry. Young Shikata was deeply interested in this work, and finally, in 1923, he joined the Heyrovský's group in Prague (41.2).

Under Professor Heyrovský, he studied the electrochemical reduction of nitrobenzene on a dropping mercury electrode. This was the first polarographic study of organic substances. During this study, Shikata discovered for the first time the phenomenon of polarographic current maximum and gave it an explanation based on the adsorption and desorption of nitrobenzene at the electrode surface.

The progress in the research with dropping mercury electrode in the Heyrovský's group was limited by the fact that the point-by-point measurements and manual plotting of the current-voltage curves were tedious and time-consuming. Considerable improvement was achieved in 1924 when the cooperation of Heyrovský and Shikata led to construction of an apparatus which registered current-voltage curves automatically (41.3). The new instrument recorded current-voltage curves (polarograms) in a few minutes, whereas manual recording took an hour or longer. In their joint paper (J. Heyrovský, M. Shikata, *Rec. Trav. Chim. Pays-Bas*, 1925, *44*, 496), Heyrovský and Shikata proposed for the instrument the name *polarograph* and for the studies of electrolysis with the dropping mercury electrode coined the term *polarography*.

The Heyrovský-Shikata instrument (41.3) consisted of a spring-driven phonograph motor, a potentiometric wheel, and a device geared to this wheel in a fixed ratio for recording galvanometer deflections (41.4). The potentiometric wheel was rotated by means of a friction drive, and thus, the voltage applied to a dropping mercury electrode was increased. Simultaneously, but at a much lower rate, a sheet of photographic paper, fixed on a drum, was moved past a slit in an otherwise light-tight box. Light reflected from the mirror galvanometer was focused on this photographic paper. Changes in current obtained during the change in voltage would then deflect the galvanometer and be recorded as a curve with the voltage abscissa and the current as the ordinate. This rather simple device operated as a new electrochemical device named polarograph.

Development of an automatic method of recording, a short time after the discovery of the new technique, played an important role in the development and dissemination of the method based on electrolysis with a dropping mercury electrode. Currently, most physico-chemical and analytical methods are carried out using automated techniques, but in the early twenties, an automated instrument represented very progressive and advanced instrumentation (e.g., recording spectrophotometers became generally accessible only two decades later).

The first commercially available polarograph based on the Heyrovský-Shikata design was produced by the company of Drs. V. and J. Nejedly in Prague, Czechoslovakia (currently Czech Republic) (41.5).

Upon returning to Japan, Shikata organized a good scientific group and carried out many outstanding works not only in polarography but also in the field of wood chemistry and related subjects. As he continued the polarographic study of organic substances, he became especially interested in clarifying the correlation of polarographic reduction potentials of organic compounds with their structures. The polarographic electronegativity rule formulated by Shikata represented the ideas resulting from the extensive studies of polarography of organic substances. Shikata also made many pioneering works in applying polarography for analysis of biological materials, food, and fermentation products.

41.1. Masuzo Shikata (photo probably taken in his office at the Department of Agricultural Chemistry, Kyoto University, in 1942)

41.2. Photo of Masuzo Shikata (*left*) and J. Heyrovský taken on December 23, 1923, in London, when Shikata and Heyrovský visited there to give papers about polarography

41.3. The first polarograph designed by Heyrovský and Shikata (1924)

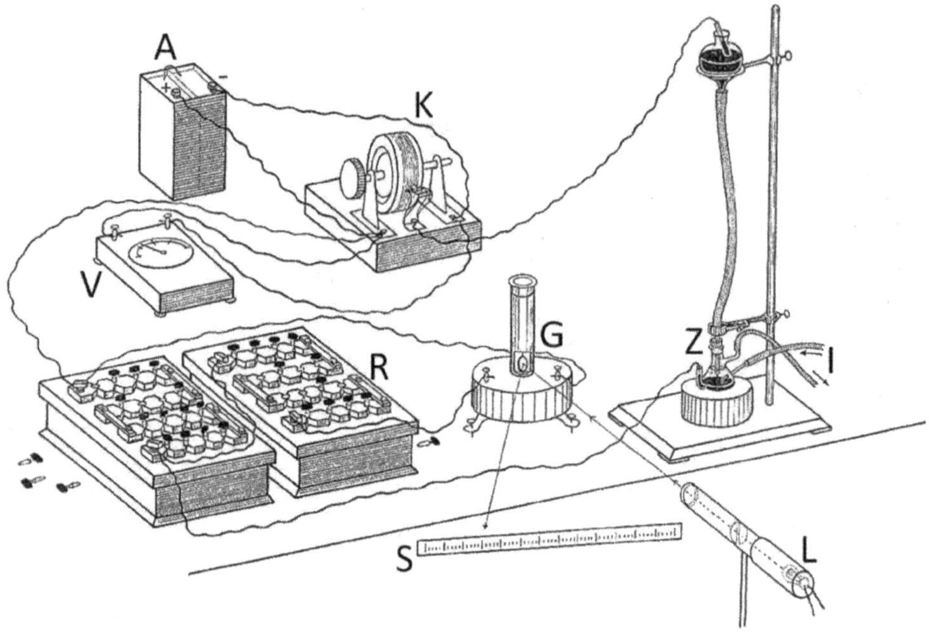

41.4. Schematics of the first Heyrovský-Shikata apparatus for measuring current-voltage curves in electrolysis with a dropping mercury electrode and photographic recording: A, battery; K, potentiometric wheel; V, voltmeter; G, mirror galvanometer; Z, electrochemical cell with a dropping mercury electrode and a counter electrode; L, source of light; S, scale (photographic paper); R, variable resistance; I, inlet and outlet for Ar or N_2 needed for deaeration. Note that the term *polarograph* was introduced only in 1925.

41.5. Early commercial model (V-301) of the Heyrovský-Shikata's photographic recording polarograph manufactured by the company of Drs. V. and J. Nejedly in Prague, Czechoslovakia

42

JOHN ALFRED VALENTINE BUTLER

Born: February 14, 1899, Winchcombe, United Kingdom
Died: July 16, 1977

John Alfred Valentine Butler (42.1) was a British physical chemist who greatly contributed to theoretical electrochemistry, particularly in connection to electrochemical kinetics and thermodynamics. The famous Butler-Volmer equation was named after him (and Max Volmer, German physical chemist, 1885–1965).

$$i = i_0 \cdot \left[\exp\left(\frac{\alpha_{\mathrm{A}} nF}{RT}\eta\right) - \exp\left(-\frac{\alpha_{\mathrm{C}} nF}{RT}\eta\right)\right]$$

This equation shows the connection between the current (i) and voltage (η) applied on the electrode. Butler was the first to connect the kinetic electrochemistry built up in the second half of the twentieth century with the thermodynamic electrochemistry that dominated the first half. Butler had, to his credit, not only the first exponential relation between current and potential (1924) but also (along with R. W. Gurney) the introduction of energy-level thinking into electrochemistry (1951). However, Butler was not alone in this study, and therefore, it is necessary to give credit also to Max Volmer, a great German surface chemist and his student (at that time) Erdey-Gruz. Butler's very early contribution in 1924 and the Erdey-Gruz and Volmer contribution in 1930 form the basis of phenomenological kinetic electrochemistry. The resulting famous Butler-Volmer equation is very important in electrochemistry.

42.1. John Alfred Valentine Butler

43

TIBOR ERDEY-GRÚZ

Born: October 27, 1902, Budapest, Austro-Hungarian Empire (presently Hungary)
Died: August 16, 1976, Budapest, Hungary

Tibor Erdey-Grúz (43.1) was a Hungarian chemist and politician. His contribution to electrochemistry was mostly in the theoretical description of electrochemical kinetics as the function of the potential applied. In most of electrochemical textbooks, the derived equation is named as the Butler-Volmer equation, giving no credit to Erdey-Grúz since he was a student when the theoretical work was published. However, some sources, particularly in Hungary, acknowledge his contribution to the theoretical study.

Erdey-Grúz performed experimental and theoretical works (43.2) on the transport processes in electrolyte solutions, electrocrystallization, adsorption, catalysis, and oxygen overpotential. However, his most important result was achieved while working with Max Volmer (German physical chemist, 1885–1965) in Berlin. They published together a pivotal paper on the hydrogen overpotential in which the proper relationship between the current and the electrode potential was formulated for the first time (Erdey-Gruz–Volmer equation). This seminal paper can be considered as a landmark since the Nernstian equilibrium theory.

The importance of their contribution has been widely acknowledged. John Bockris (South African/American electrochemist, 1923–2013) expressed this view in a most adequate manner: "Erdey-Gruz and Volmer, whom I see as the fathers of electrode kinetics…" Later, Erdey-Gruz published several books on transport processes and electrochemical kinetics.

Notably, the similar work, performed independently, was published by John Alfred Valentine Butler (British physical chemist, 1899–1977). Then the final version of the equation is usually named as Butler-Volmer equation, giving no credit to Erdey-Grúz.

$$i = i_0 \cdot \left[\exp\left(\frac{\alpha_A nF}{RT}\eta\right) - \exp\left(-\frac{\alpha_C nF}{RT}\eta\right)\right]$$

where

- i as current density, A/m^2;
- i_0 as exchange current density, A/m^2;
- η as overpotential applied;
- α_A and α_C as charge transfer coefficient, anodic and cathodic, respectively;
- F as Faraday constant;
- R as universal gas constant;

- T as absolute temperature, K; and
- n as number of electrons involved in the electrode reaction.

The transfer coefficient (α) introduced by Erdey-Gruz and Volmer permits the interpretation of the experimental results without a detailed knowledge of the transition state (activated complex) and the structure of the double layer at the electrode interface.

Overall, Butler's very early contribution in 1924 and the Erdey-Gruz and Volmer contribution in 1930 form the basis of phenomenological kinetic electrochemistry. The resulting famous Erdey-Gruz-Butler-Volmer equation is very important in electrochemistry.

43.1. Tibor Erdey-Grúz

43.2. Tibor Erdey-Grúz working in his laboratory (1930s)

44

WIKTOR KEMULA

Born: March 6, 1902, Ismail, Bessarabia, Russian Empire (now Ukraine)
Died: October 17, 1985, Warsaw, Poland

Wiktor Kemula (44.1) was a Polish electrochemist who greatly contributed to the development of electroanalytical chemistry, particularly polarography. He is known for inventing chromatopolarography and developing the hanging mercury drop electrode (HMDE).

Kemula's scientific interest was mostly concentrated on electroanalytical chemistry, particularly on polarography (electrochemical measurements using a dropping mercury electrode invented by Jaroslav Heyrovský, a Czech electrochemist, 1890–1967). It should be noted that polarography was almost the only electroanalytical method known in the first half of the twentieth century.

Kemula invented chromatopolarography, a combination of chromatography with polarographic detection. The eluent leaving a chromatography column entered a polarographic cell, and the consecutive analytes were determined amperometrically. It should be noted that the term *chromatopolarography* refers to the use of a mercury-dropping electrode, which is not in use presently. Then this term is presently changed to *amperometric* or *electrochemical* chromatography, where different solid metal electrodes are used instead of mercury. Overall, the chromatopolarography invented by Kemula was a precursor of modern high-performance liquid chromatography (HPLC) and electrophoresis with the electrochemical detection of the analytes.

Another important invention made by Kemula was a hanging mercury drop electrode (HMDE), where the electrochemical analysis was performed using a single Hg drop (44.2A). The HMDE produces a mercury drop of controlled geometry and surface area at the end of a capillary in contrast to the dropping mercury electrode (DME) (44.2B), which steadily releases drops of mercury during an experiment.

Notably, the Hg drops released from a capillary of the DME are constantly changing their size and shape until they become too heavy and drop down. This process results in the increasing capacitance (background) current, affecting the detection limit. The HMDE does not have this problem and allows higher sensitivity. The HMDE allows simple substitution of the used Hg drop with a new fresh droplet that is advantage compared to solid metal electrodes, which require cleaning of their surfaces.

44.1. Wiktor Kemula

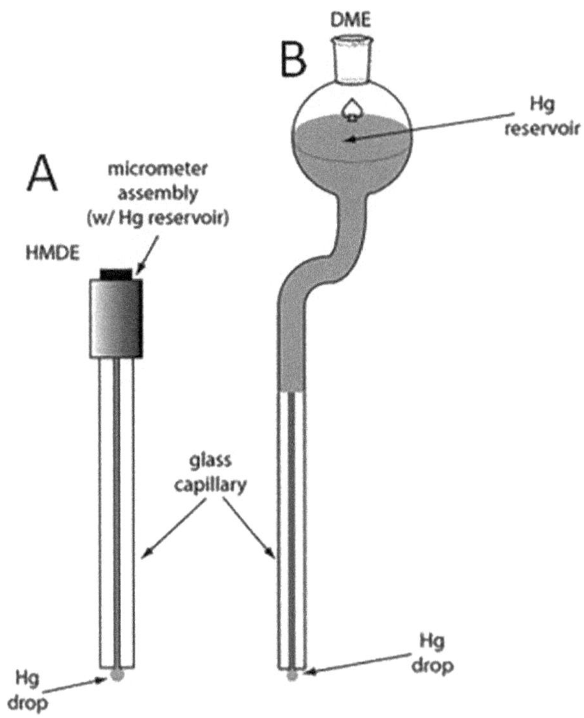

44.2. Comparison of the HMDE (A) with DME (B)

45

FRANCIS THOMAS BACON

Born: December 21, 1904, Billericay, Essex, United Kingdom
Died: May 24, 1992, Little Shelford, near Cambridge, United Kingdom

Francis Thomas Bacon (45.1) was an English mechanical and electrical engineer mostly known for construction of practically useful fuel cells, particularly used in the US-Apollo space program.

Fuel cells were first demonstrated by Sir William Robert Grove (British scientist, 1811–1896) in 1839. (Actually, the fuel cell concept was demonstrated much earlier by Humphry Davy and Christian Friedrich Schönbein.) However, his invention was unused for over a hundred years until it was revived by Bacon. Bacon initiated his work on the hydrogen-oxygen (H_2-O_2) fuel cell in 1932 when he constructed the first efficient fuel cells (45.2). In this kind of fuel cells, H_2-gas was used as a fuel and O_2-gas was an oxidizer, working together to produce electrical power. Bacon replaced acid-electrolyte solution with a less corrosive potassium hydroxide (alkaline) electrolyte. This allowed the use of less expensive nickel electrodes instead of platinum electrodes originally used by Grove.

Another important improvement was the invention of the porous gas diffusion electrodes, providing a high active surface area, thus significantly increasing the generated current. The anode- and cathode-electrolyte solutions were separated with an ion-exchange membrane. Bacon's alkaline fuel cells operated between 100°C and 250°C, while presently, this kind of fuel cells operate around 70°C, being very convenient for their practical applications at a relatively low temperature.

The most famous use of the Bacon fuel cell was electrical power generation in moon flights of the Apollo space program (45.3). The Apollo program was the United States human spaceflight program carried out by the National Aeronautics and Space Administration (NASA), which succeeded in landing the first men on the moon from 1968 to 1972. Notably, the fuel cell operation resulted in production of water as the result of H_2 oxidation. This water was used by the astronauts during the space mission.

45.1. Francis Thomas Bacon

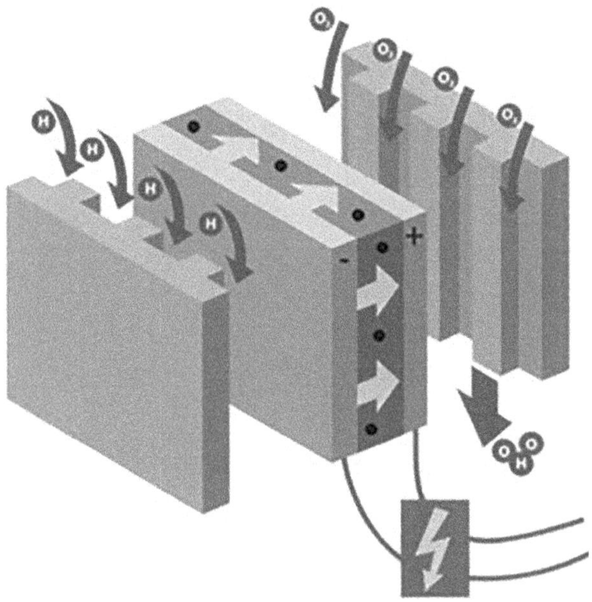

45.2. Schematics of the H_2-O_2-gas fuel cell constructed by Bacon in 1932

45.3. A fuel cell used in the Apollo spacecraft service module

46

MARCEL POURBAIX

Born: September 16, 1904, Tula, Russian Empire (presently Russia)
Died: September 28, 1998, Uccle, Belgium

Marcel Pourbaix (46.1) was a Belgian chemist (born in Russia) who greatly contributed to studies on corrosion. His biggest achievement is the derivation of potential-pH diagrams, better known as Pourbaix diagrams (46.2A). Pourbaix diagrams are thermodynamic charts constructed using the Nernst equation. They visualize the relationship between possible redox states of a system, bounded by lines representing the reactions between them under thermodynamic equilibrium. The Pourbaix diagrams can be read much like phase diagrams. In 1963, Pourbaix produced *Atlas of Electrochemical Equilibria in Aqueous Solutions* (46.2B), which contains potential-pH diagrams for all elements known at the time. Pourbaix and his collaborators began preparing the work in the early 1950s and continued the diagram updates over many years.

46.1. Marcel Pourbaix

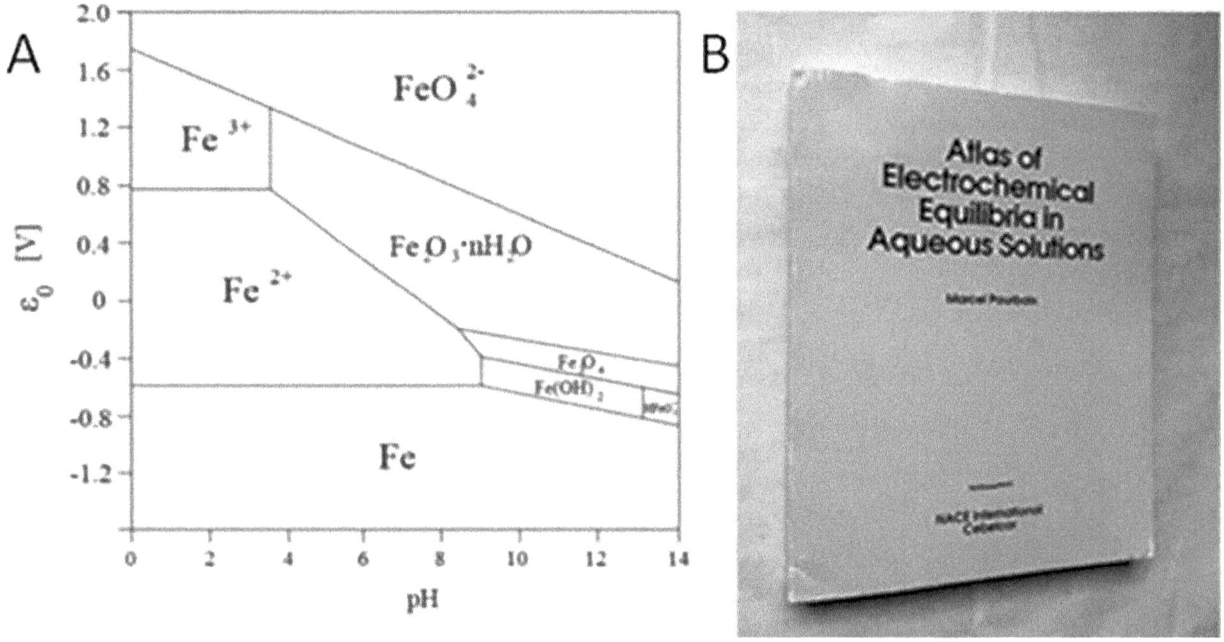

46.2. (A) Pourbaix Diagram for Fe compounds. (B) *Atlas of Electrochemical Equilibria in Aqueous Solutions*.

47

RUDOLF BRDIČKA

Born: February 25, 1906, Praha, Austro-Hungarian Empire, (presently Czech Republic)
Died: June, 25, 1970, Mariánské Lázně, Czechoslovakia (presently Czech Republic)

Rudolf Brdička (47.1) was a Czech physical chemist and electrochemist, particularly known for his research on biomedical applications of polarography.

Brdička was a pupil and later a collaborator of Prof. Jaroslav Heyrovský (the inventor of the polarographic method and recipient of the Nobel Prize in 1959). Brdička, following his studies on polarography performed with Heyrovský, devoted all his scientific career to the use of polarography for different electroanalytical applications. Since at that time polarography was used for electrochemical analysis of small organic redox molecules cations and their complexes, Brdička studied the electrochemistry of cobalt cations (Co^{3+}).

While the Co^{3+} polarographic wave was following the expected redox behavior similar to other polarographic metal ion reactions, surprisingly, very unusual polarographic waves were observed in the presence of some proteins. The observed phenomenon was explained as a catalytic redox process that involves complex formation between Co^{3+} cations with thiol (-SH) groups in the protein backbone. The polarographic waves were named Brdička waves. The exact mechanism, which involves two catalytic processes proceeding at different potentials, was elucidated in detail later. It was shown that the electrochemical process includes the redox process of the thiol-complex of Co^{3+} and then catalytic reduction of H^+ cations and H_2 evolution at more negative potentials, thus resulting in double polarographic waves. The observed waves were used as a very sensitive indication of proteins (note that it was a catalytic process), and the waves were specific to different kinds of proteins (note that they were dependent on the presence of thiol groups in the proteins). The Brdička waves were used in the analysis of protein-biomarkers of cancer and other health problems over several decades (47.2).

Presently, the polarographic analysis is not used, and Brdička waves have only historic interest. Notably, Brdička waves originate from the redox processes of thiol groups in the peripheral cysteine residues. Thus, they are not related to the redox transformations of enzyme active centers, which are important for various modern biosensor and bioelectronic applications.

47.1. Rudolf Brdička

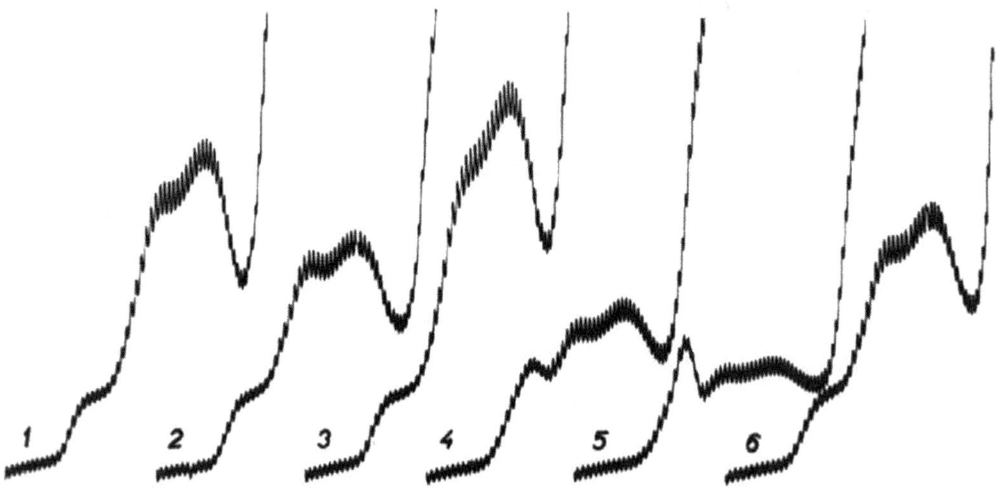

47.2. Polarographic Brdička waves obtained with human serum samples from patients with (1) status febrilis, (2) tumor hepatitis, (3) *Sarcina ventriculi*, (4) normal serum, (5) hepatic cirrhosis, and (6) atherosclerosis

48

TATYANA ALEKSANDROVNA KRYUKOVA

Born: January 29, 1906, Moscow, Russian Empire (now Russia)
Died: September 7, 1987, Moscow, USSR (now Russia)

Tatyana Alexandrovna Kryukova (48.1), a Russian scientist and electrochemist, made important contributions to electroanalytical chemistry (48.2). She particularly worked in close collaboration with Professor Aleksandr Naumovich Frumkin, who was the greatest Russian scientist in the area of electrochemistry. Kryukova is particularly remembered for developing the theory of polarographic maxima, which were observed as a sharp increase in the current produced upon polarographic measurements under some conditions (48.3). These current peaks originated from tangential movements (rotation) of a droplet mercury electrode, stimulating diffusion in the depletion layer and current increase.

Kryukova experimentally observed and theoretically explained the formation and then inhibition of these peaks upon adsorption of organic substances (mostly surfactants) on a mercury droplet electrode. It should be noted that for the first time, the effect of surfactants on polarographic measurements was reported in the 1920s in the laboratory of Prof. Jaroslav Heyrovský (polarography inventor and Nobel Prize laureate in 1959), and the study of this effect was published in 1931. However, the study of the surfactant effect performed by Heyrovský was only fragmental. The credit for a detailed explanation of the reasons for the polarographic maxima origin and a systematic study of this effect belongs to Kryukova.

In 1949, Kryukova discovered another very unusual phenomenon, later named as Kryukova effect (48.4). This effect was observed as a sudden decrease in the current at very negative potentials upon polarographic reduction of anionic species—for example, persulfate or dichromate anions—particularly when a very diluted supporting electrolyte was present in the analyte solution. This current minimum disappeared when the electrolyte concentration was increased.

Later in 1952, Frumkin and G. M. Florianovich (a graduate student at that time) theoretically explained the effect observed by Kryukova as the repulsion of redox anions from the negatively charged electrode surface as predicted by Frumkin's theory of 1933. This is exactly why the effect was only observed for anionic redox species particularly with very negative potentials, providing a negative charge at the working electrode. As expected, the high concentration of the supporting electrolyte was screening the electrostatic interaction between the negative Hg droplet electrode and the negative anion redox species, eliminating the current decrease.

It should be noted that the electrochemical study of persulfate ions when the Kryukova effect was observed had not only gained theoretical interest, demonstrating a fundamental electrostatic effect at polarized electrodes, but it was also practically important as a part of the Russian uranium project because persulfate ions were used as a reagent in the separation of uranium isotopes.

Kryukova published many important research papers, and her 800-page textbook *Polarographic Analysis* written in collaboration with a few other scientists and published in Russian in 1959 was one of the most important educational materials and a reference guide on polarography in the USSR for many years.

48.1. Tatyana Kryukova (1940s)

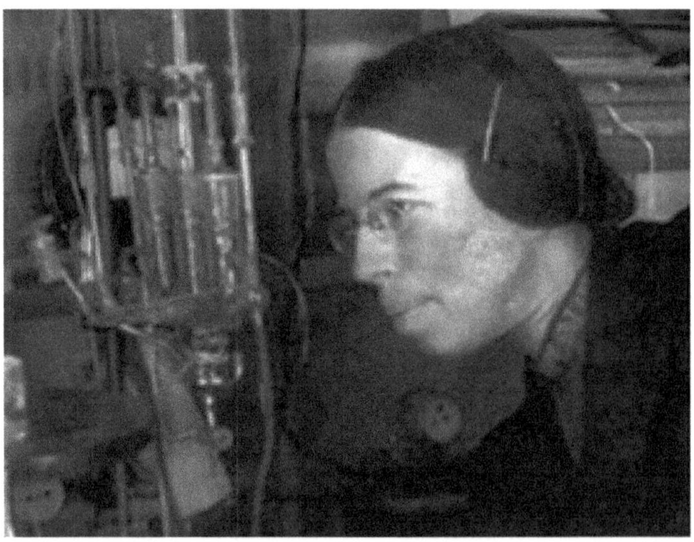

48.2. Professor Tatyana Kryukova performing a polarographic experiment

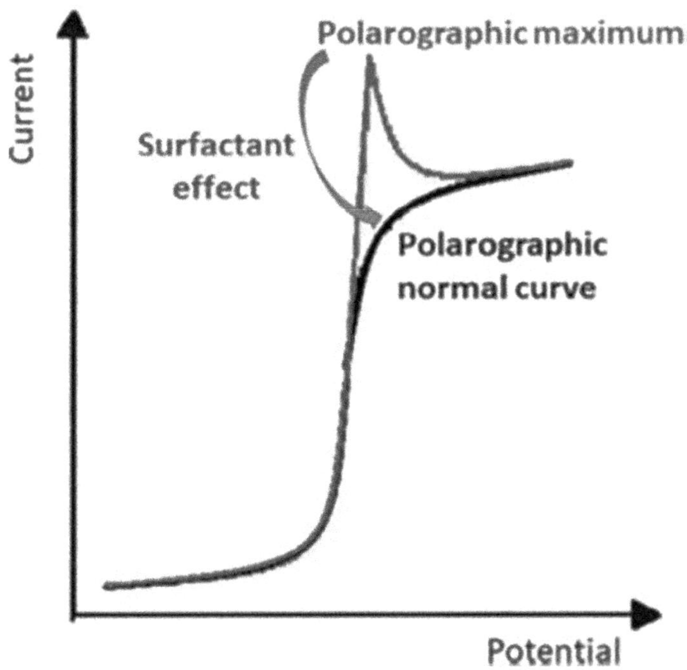

48.3. Polarographic maximum and its suppression in the presence of a surfactant

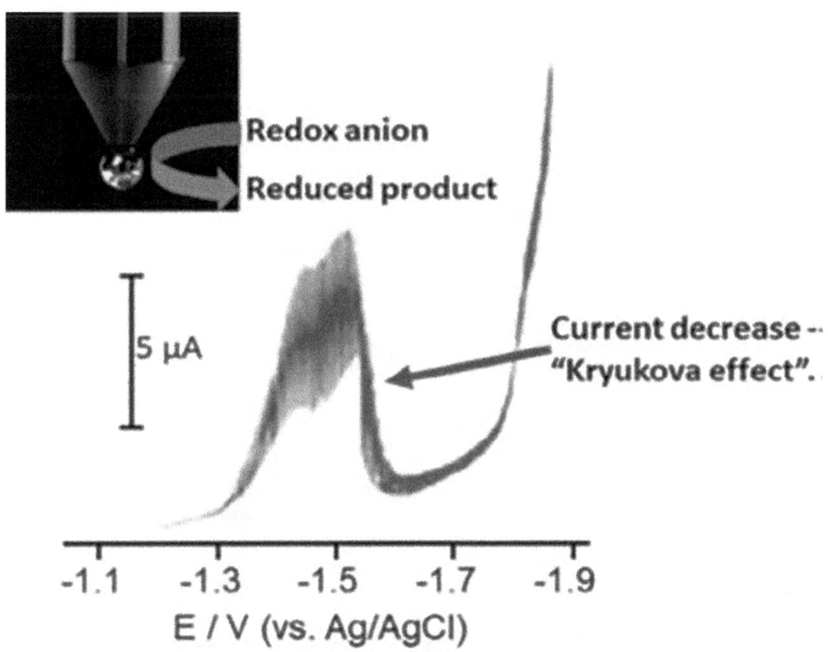

48.4. A polarographic curve with the Kryukova effect. The inset shows a mercury-dropping electrode used for the electrochemical reduction of anion redox species, leading to the current decrease upon reaching very negative potentials applied on the electrode.

49

YAKOV PEYSACHOVICH GOKHSHTEIN

Born: September 23, 1906, Golendry, Russia Empire (now Ukraine)
Died: November 29, 1996, Moscow, Russia

Yakov Peysachovich Gokhshtein (49.1) was a Russian/Soviet (with Jewish ethnicity) electrochemist.

Gokhshtein received PhD in 1935 for his work on polarography. Note that polarography was a very new technique at the time it was just invented in 1922 by Jaroslav Heyrovský (Czech chemist and inventor, 1890–1967). Then he continued the polarographic studies that were quite typical for his time. Note that polarography was the only available electroanalytical method until 1950s–1960s. In his polarographic studies, Gokhshtein used various methods—for example, oscillographic polarography (oscillo-polarography).

In 1961, Yakov Gokhshtein invented, with his son Alexander, the first automatically working static mercury drop electrode. The static mercury drop electrode (SMDE) was a significantly new electrode for its use in polarography. The electrode had two distinct operating regimes: a drop growth period followed by a stationary period with a mercury droplet keeping its size unchanging, then minimizing the capacitance current resulting in increased sensitivity.

Gokhshtein has also contributed to the theory of diffusion-controlled electrochemical reactions.

49.1. Yakov Peysachovich Gokhshtein

50

DIONÝZ ILKOVIČ

Born: January 18, 1907, Šarišský Štiavnik, Austria-Hungary Empire (presently Slovakia)
Died: August 3, 1980, Bratislava, Czechoslovakia (presently Slovakia)

Dionýz Ilkovič (50.1) was a Czechoslovak physicist and physical chemist. He made fundamental contributions to the theoretical background of polarography and electroanalytical chemistry in general.

Polarography is the first electroanalytical technique that performs a voltammetric study with a dropping mercury electrode (50.2A). This technique was invented in 1922 by Czech physical chemist Jaroslav Heyrovský, who received the Nobel prize in 1959 for the polarography invention and its application to numerous electroanalytical studies. The polarography made the background for different electroanalytical methods, particularly cyclic voltammetry, and other modern voltammetric techniques.

Ilkovič was a research assistant for Professor Heyrovský in Prague, being one of the cofounders of polarography. He developed a theoretical description of the polarographic measurements considering the current based on the diffusion of the redox species (50.2B). The mathematic expression was named after him as Ilkovic equation: $i_d = k\ n\ c\ m^{2/3}\ t^{1/6}\ D^{1/2}$, where i_d is the diffusion-limited polarographic current, c is the concentration of the redox species, D is the diffusion coefficient of the redox species, n is the number of electrons involved in the interfacial redox process, m is the flow rate of the mercury from the capillary, t is the mercury drop lifetime, and k is the constant coefficient, which includes the Faraday constant and the density of mercury.

The Ilkovic equation was highly important for the quantitative analysis of the polarographic measurements.

50.1. Dionýz Ilkovič

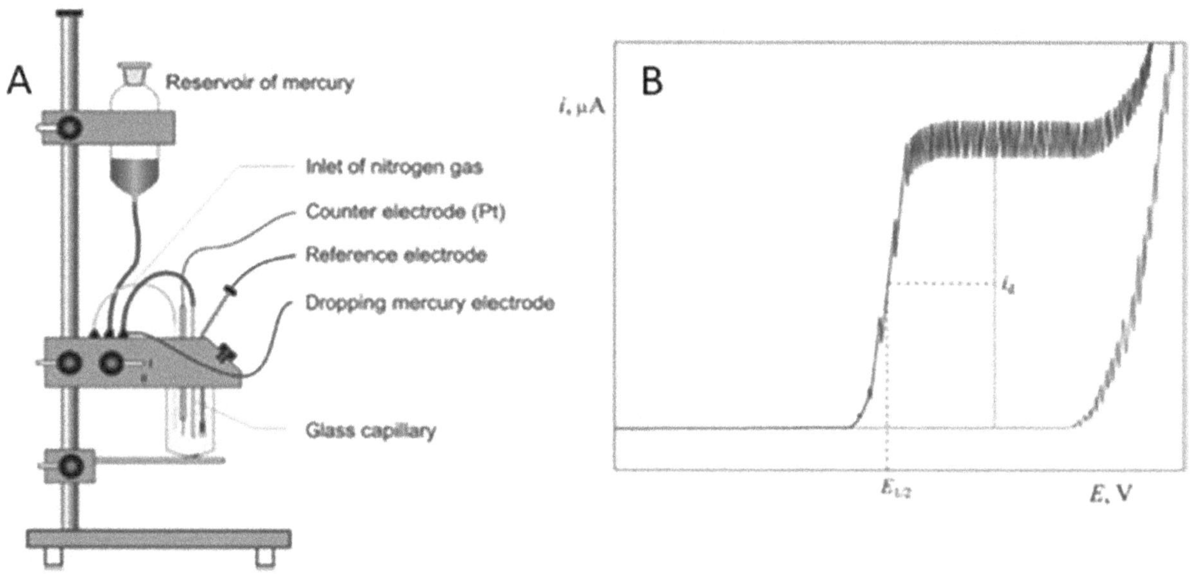

50.2. (A) Schematics of the polarographic setup for the voltammetric measurements on a dropping mercury electrode. Note that the first polarography experiments were performed with two electrodes (working and counter) only. The third reference electrode became used much later when potentiostats started to be available. (B) An example of the polarographic curves. The horizontal curve shows the background polarogram recorded in the absence of the redox species. The oscillating curve shows the polarogram in the presence of the added redox species. The polarographic wave corresponds to the reduction of the redox species (cathodic current).

51

JAMES J. LINGANE

Born: September 13, 1909, St. Paul, Minnesota, USA
Died: March 14, 1994, Lexington, Massachusetts, USA

James J. Lingane (51.1) was an American electrochemist, mostly working in the field of polarography. This electroanalytical technique—invented in 1922 by Jaroslav Heyrovský (Czech electrochemist, awarded with the Nobel Prize, 1890–1967)—was a very new technique used that time for analytical applications.

While working with Izaak Maurits Kolthoff (American analytical chemist, 1894–1993) as a PhD student, Lingane was inspired by Heyrovský's work on polarography, particularly after meeting him personally during Heyrovský's visit to the USA in 1933. Polarography became Lingane's favorite research area from his PhD study and for the whole of his life. Soon, Lingane became a leader in the USA in the field of polarography and some other electroanalytical methods, such as controlled-potential electrolysis, coulometry, coulometric titration, etc.

These novel methods required new electrical/electronic instrumentation. Lingane took the challenge to design such instruments. Particularly, he constructed an early electromechanical Lingane-Jones potentiostat (51.2A). Notably, Lingane greatly contributed to the construction of polarographs (instruments for polarographic studies) based on the Heyrovský-Shikata concept but with improved operation. He designed a simple electric circuitry to realize a polarographic instrument, which was capable of higher precision and accuracy than most commercial instruments of that time. This was one of the simplest circuits, which satisfied all requirements for the electroanalytical applications.

Another example of the instruments constructed by Lingane was a multipurpose electroanalytical instrument (51.2B). This instrument was operating with various functions, including the potentiostat function, maintaining the potential of a working electrode constant during electrolysis, electrogravimetric analysis of metals, electrolytic separations of metals prior to final determinations by other methods, coulometric analysis, and electrolytic preparations by the controlled-potential technique. Electrolysis—at constant total applied voltage or with constant current (i.e., operating in potentiostatic or galvanostatic regimes)—was also possible.

In 1953, Lingane wrote the widely used monograph *Electroanalytical Chemistry*, which was used for several decades as the most comprehensive textbook. Another famous book on polarography was written in coauthorship with Kolthoff in 1952 (51.3). This book was a very important contribution, along with other books on polarography written by Heyrovský, for dissemination of polarography as an electroanalytical method.

51.1. James J. Lingane

51.2. (A) Lingane-Jones potentiostat. (B) Lingane multipurpose electrochemical servo instrument.

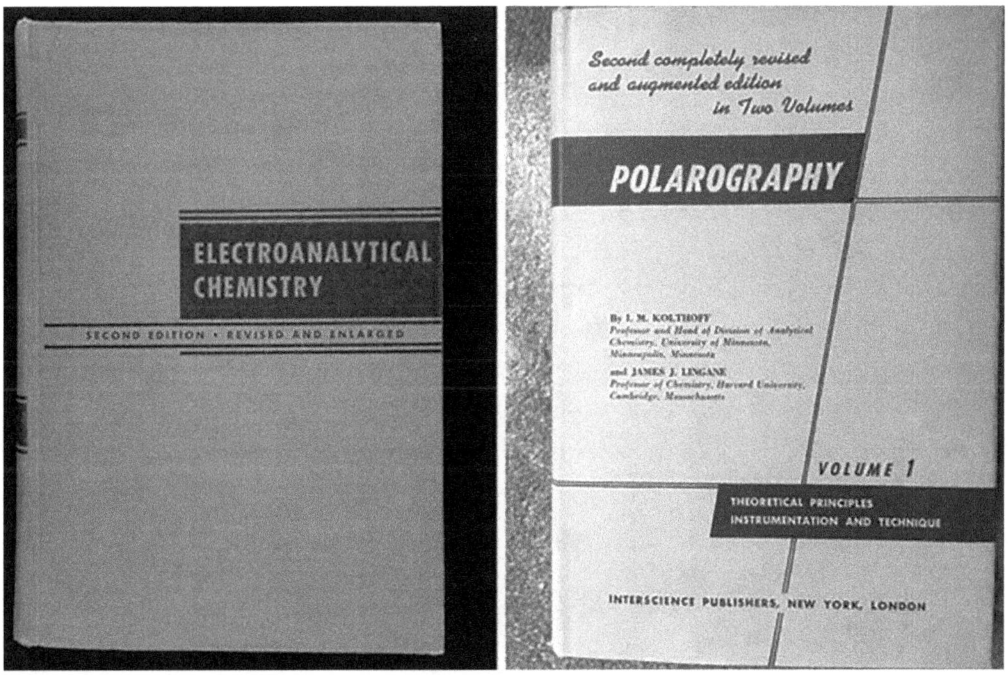

51.3. Lingane's books: J. J. Lingane, *Electroanalytical Chemistry*, first edition, 1953, and second edition, 1958, Interscience Publishers (*left*). I. M. Kolthoff and J.J. Lingane, *Polarography*, 2 volumes, first edition, 1952, and second edition, 1965, Interscience Publishers (*right*)

52

JOHN EDWARD BROUGH RANDLES

Born: 1912

Died: 1998

John Edward Brough Randles (52.1) was an English electrochemist who made important contributions to the theoretical background of polarography, cyclic voltammetry, and electrochemical impedance spectroscopy. Many modern techniques of electrochemistry are descended from his work, including cyclic voltammetry, anodic stripping voltammetry, and various types of hydrodynamic voltammetry. Randles-Ševčík equation applied on linear sweep voltammetry and cyclic voltammetry and Randles equivalent circuit used in the modeling of impedance spectra are named after him.

The earliest electrochemical work of Randles performed with an oscillopolarograph (cathode ray polarograph) resulted in the development of linear sweep voltammetry. In addition to the experimental work, Randles solved a theoretical problem for expressing the current for diffusion-controlled electrochemical reactions by applying an ingenious graphical method.

His theoretical work resulted in the equation for the peak current later named as Randles-Ševčík equation. The second contributor working on this theoretical problem, Ševčík, was the Czech scientist who obtained a similar equation using a less precise numerical approach. The derived equation is given by:

$$i_p = 0.4463\, nFAC \sqrt{\frac{nFvD}{RT}}$$

where i_p is current maximum in amps; n is number of electrons transferred in the redox event; A is electrode area in cm^2; F is Faraday Constant in C mol^{-1}; D is diffusion coefficient in cm^2/s; C is concentration in mol/cm^3; v is scan rate in V/s; R is Gas constant in J K^{-1} mol^{-1}; T = temperature in K.

Another important contribution of Randles to electrochemistry was in the theoretical analysis of Faradaic impedance spectra published in 1947. The Randles equivalent circuit has been applied to the analysis of the impedance spectra, including interfacial electron transfer (Faradaic component), capacitance, and diffusion contributions to the impedance. It became the most frequently used theoretical treatment of impedance spectra. It should be noted that similar results were obtained by Russian scientists Dolin and Erschler in 1940, but the papers published in the Russian language have not been seen by the electrochemical community.

The Randles equivalent circuit (52.2) is one of the simplest and most common circuit models of electrochemical impedance. It includes a solution resistance, a double-layer capacitor, and a charge transfer or polarization resistance. While the Randles equivalent circuit is frequently sufficient for modeling simple electrochemical systems, it can be used as a starting point for more sophisticated models—for example, those based on more resistances and capacitances organized in parallel or in a sequence.

It should be noted that John Randles was not only working on solving theoretical problems in electrochemistry, but he was a very good experimentalist. As an example of his experimental work, the Volta potential difference between a mercury droplet and an electrolyte solution measured by Randles can be mentioned. Notably, the great Russian electrochemist Alexander Frumkin had failed to obtain a stable and reproducible result for this kind of measurement.

Randles published relatively few papers, but many of them are of great importance, and his theoretical treatments of electrochemical systems have been included in all electrochemistry textbooks.

52.1. John Randles

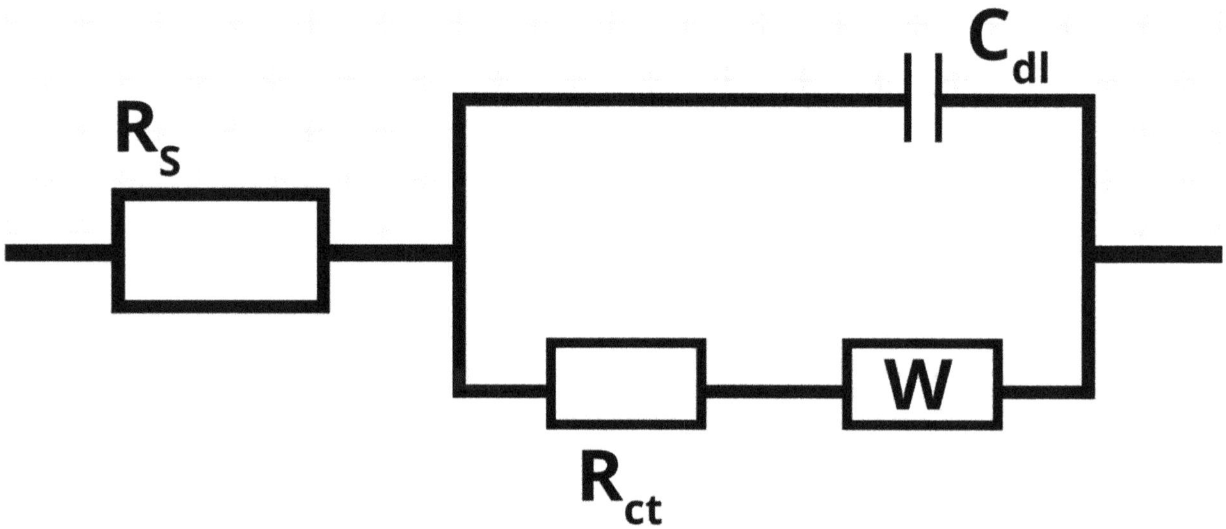

52.2. The Randles equivalent circuit: R_s is an Ohmic resistance of the electrolyte solution, C_{dl} is a double layer capacitance, R_{ct} is a charge transfer resistance, and W is a Warburg impedance.

53

GEOFFREY CECIL BARKER

Born: October, 2, 1915, Belper, Derbyshire, United Kingdom
Died: March 31, 2000, Oxford, United Kingdom

Geoffrey Cecil Barker (53.1) was an English electrochemist who is remembered for invention of novel polarographic techniques, particularly important for electrochemical analysis with extremely high sensitivity.

Polarography electroanalytical method invented by Jaroslav Heyrovský (Czech electrochemist, 1890–1867) was originally based on direct current (DC) measurements using a dropping mercury electrode as a working electrode. This method was a revolutionary approach to electroanalytical studies, being a precursor of all presently used electroanalytical techniques, mostly for broadly used cyclic voltammetry. Heyrovský's polarography invention and its use for various electroanalytical applications resulted in 1959 Nobel Prize in Chemistry.

Geoffrey Barker's contribution was the invention of pulse current and alternative current (AC) measurements with a dropping mercury electrode. The methods invented by Barker provided extremely high sensitivity, not achievable with the original DC measurements. In 1952, he invented square wave polarography, and in 1958, he invented normal and differential pulse polarography and radio frequency polarography.

Normal polarography invented by Heyrovský has been replaced by various forms of pulse polarography (53.2). Normal pulse polarography (53.2A) uses a series of potential pulses. The current is sampled at the end of each potential pulse before returning the potential to its original value. The shape of the resulting voltammogram is similar to that of the normal polarography but without the current oscillations related to the formation and increase of the mercury droplet.

Since the potential is applied for a small fraction of the Hg droplet lifetime, the redox analyte undergoes oxidation or reduction for a short period of time, thus resulting in a smaller diffusion layer. As a result, the faradaic current in normal pulse polarography is greater than in classical polarography, resulting in better sensitivity and smaller detection limits. In differential pulse polarography (53.2B), the current is measured twice per cycle: before applying the pulse and at the end of the cycle. The difference in the two currents gives rise to the peak-shaped voltammogram with significantly increasing sensitivity and decreasing the detection limit by two orders of magnitude. Other forms of pulse polarography include staircase polarography (53.2C) and square wave polarography (53.2D). The advantage of square-wave polarography is a much shorter time of the potential application, then significantly decreasing analysis time. This can allow the whole detection time within the lifetime of a single Hg droplet.

Barker applied for the first time faradaic rectification to electroanalytical measurements. Besides these techniques, Barker developed several other variants of polarography. Barker's inventions led the way to the use of electronic control in electroanalytical instrumentation.

The high-sensitivity polarographic techniques invented by Barker allowed trace analysis of heavy metals in seawater. Differential pulse polarography appeared to be suitable for DNA analysis, allowing the determination of minor variations in the DNA structure. Later, the techniques invented by Barker became standard electroanalytical methods available with all modern electrochemical instruments and used with various solid electrodes excluding the use of mercury.

53.1. Geoffrey Cecil Barker

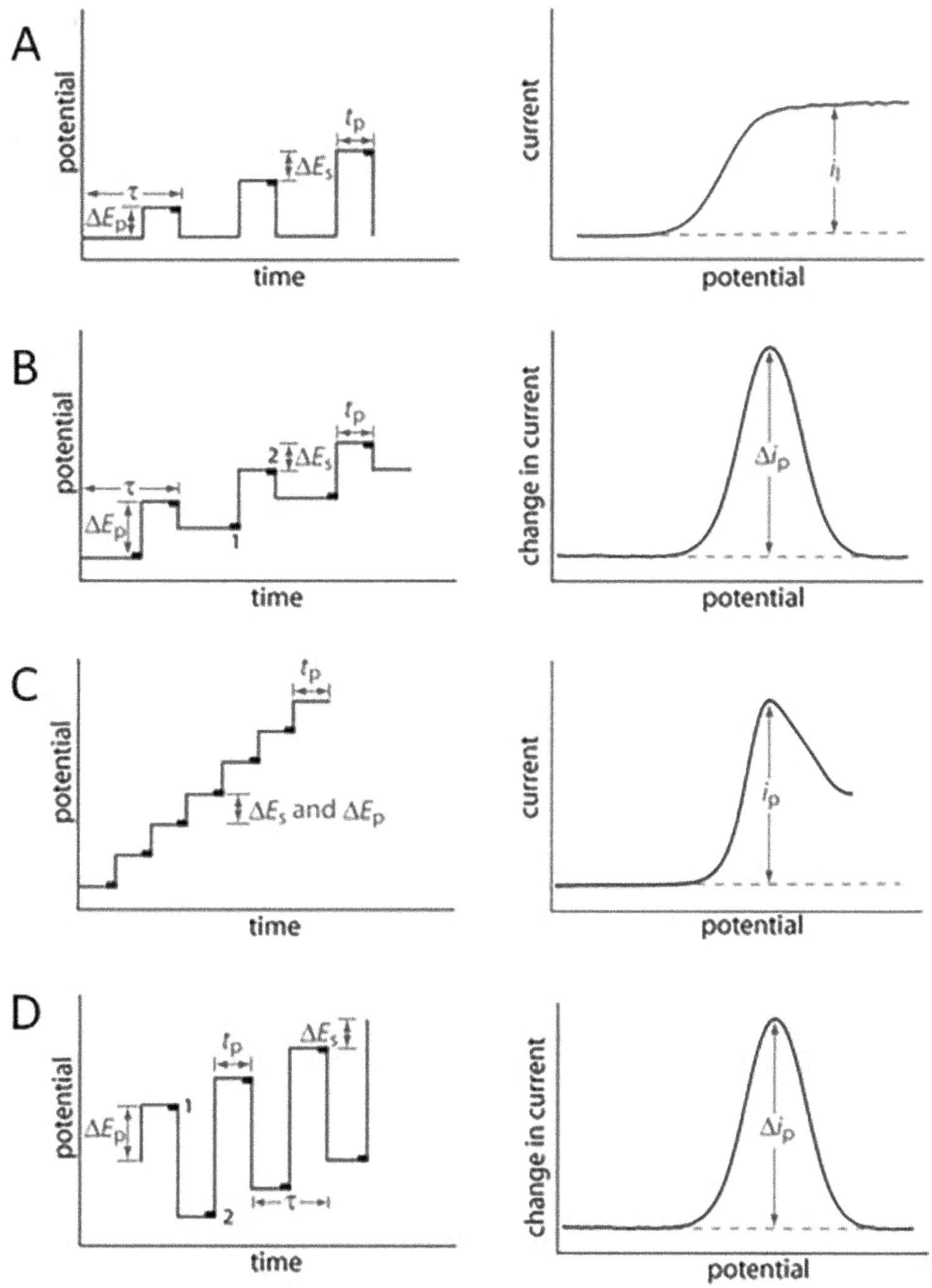

53.2. Different kinds of potential pulse/step polarography: (A) normal pulse polarography, (B) differential pulse polarography, (C) staircase polarography, (D) square wave polarography. The diagrams (*left*) show the potential application and current sampling modes: τ is the cycle time, t_p is the pulse time, ΔE_p is the pulse potential, ΔE_s is change in the potential per cycle. The polarograms (*right*) correspond to the potential application modes: i_p is the peak current, and Δi_p is the peak current difference.

54

MIRKO KALOUSEK

Born: June 19, 1915, Nakhchivan, Russian Empire (now Naxcivan Autonomous Republic, part of Azerbaijan)
Died: August 8, 1996, Prague, Czech Republic

Mirko Kalousek (54.1) was a Czech electrochemist mostly known for his contribution to the polarographic electroanalytical technique. He constructed a very convenient electrochemical cell for polarographic measurements and invented an electrical switch (commutator) used in pulse polarography.

The main feature of the polarographic cell constructed by Kalousek was the separation of a reference electrode with an electrolyte bridge using a glass cock (54.2). In his dissertation (1939), Kalousek concentrated on the reversibility of the electrochemical reduction of various metal cations at a dropping mercury electrode, followed by oxidation of the produced amalgams. For the rapid switching between the reductive and oxidative potentials, Kalousek invented a switch (commutator) bearing his name—Kalousek commutator. This commutator was based on the rapidly rotating switch that changed the potential applied on a mercury-dropping electrode with a constant controlled frequency, and the current was recorded during the higher potential half-cycles. Note that this technique was used many years before computerized electrochemical analyzers became available.

While the original motivation was to study the reduction-oxidation of metal cations, later, the method was adopted for the analysis of the reversibility of electrochemical reactions in general. At more negative potential, the reactant was reduced, and then, when the potential was switched to a positive value, the oxidation of the reduced product was possible if the process was reversible. Further modification of the method resulted in the invention of pulse polarography, particularly square wave voltammetry, providing higher sensitivity compared to the classical polarography invented by Jaroslav Heyrovský (Czech electrochemist, 1890–1967).

54.1. Mirko Kalousek

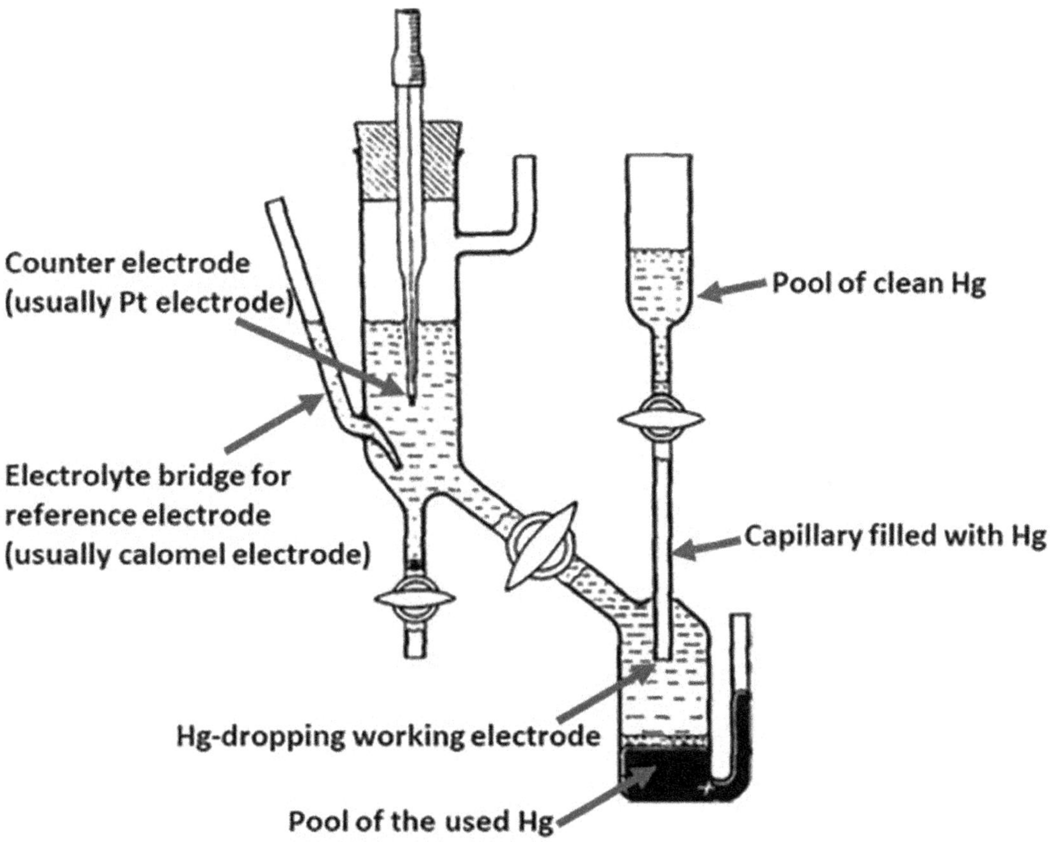

54.2. The electrochemical cell constructed by Kalousek for polarographic measurements

55

KLAUS JURGEN VETTER

Born: July 13, 1916, Berlin, Germany
Died: December 12, 1974, Berlin, Germany

Klaus Jurgen Vetter (55.1) was a German electrochemist who published almost a hundred papers on the fundamentals of electrode kinetics, overpotentials, redox reactions, ion transfer, iron passivity and corrosion, and other electrochemical topics. However, most important is his book: K. J. Vetter, *Elektrochemische Kinetik*, Springer, Berlin, 1961 (in German); K. J. Vetter, *Electrochemical Kinetics*. Academic Press, New York, 1967 (English ed.) (55.2). Note that the Russian translation was also published in 1967.

The book offers information on electrochemical thermodynamics and the theory of overvoltage. The topics covered in the book include equilibrium potentials, electrical double layer, electrocapillarity, charge transfer, diffusion, and reaction overvoltage. Crystallization overvoltage, total overvoltage, and resistance polarization are also discussed in the book. The book also covers the methods of determining electrochemical reaction mechanisms, including examination of the overall electrode reaction and determination of the type of overvoltage and reaction kinetics. Despite the fact that this book was published long time ago, it is still the best and the most comprehensive summary on electrochemical kinetics. The book offers a well-explained theoretical background and experimental data on electrochemical kinetics.

Vetter succeeded in introducing the kinetic description of complex reactions in the field of electrochemistry. He gave groundbreaking interpretations of the exchange current density, the electrochemical reaction order, and the Flade potential (related to the corrosion process).

55.1. Klaus Jurgen Vetter

ELEKTROCHEMISCHE KINETIK

VON

KLAUS J. VETTER

Springer-Verlag Berlin Heidelberg GmbH

1961

ELECTROCHEMICAL KINETICS

THEORETICAL ASPECTS

Sections 1, 2, and 3 of Electrochemical Kinetics: Theoretical and Experimental Aspects

KLAUS J. VETTER

Professor of Physical Chemistry
Director of the Institute of Physical Chemistry
Free University of Berlin

Translated by SCRIPTA TECHNICA, INC.

TRANSLATION EDITORS:

STANLEY BRUCKENSTEIN
Division of Analytical Chemistry
University of Minnesota
Minneapolis, Minnesota

BRIAN HOWARD
Department of Chemistry
New York University
New York, New York

1967

ACADEMIC PRESS • New York • London

55.2. K. J. Vetter, *Elektrochemische Kinetik*, Springer, Berlin, 1961 (in German); and K. J. Vetter, *Electrochemical Kinetics*, Academic Press, New York, 1967 (English edition)

56

VENIAMIN (BENJAMIN) GRIGORIEVICH LEVICH

Born: March 30, 1917, Kharkov, Russia, (now Kharkiv, Ukraine)
Died: January 19, 1987, Englewood, New Jersey, USA

Benjamin Levich (Veniamin Grigorievich Levich—Russian name) (56.1) was a Russian (with Jewish ethnicity) physical chemist and electrochemist. He invented electrochemical hydrodynamics—a novel science area combining electrochemical kinetics with mass transport processes. Levich provided theoretical treatments for electrochemistry at a rotating disk electrode (RDE) and derived the equation (named after him Levich equation) to find rate constants for electrochemical processes.

The novel scientific area, electrochemical hydrodynamics, invented by Levich, was particularly powerful for the analysis of electrochemical kinetics using a rotating disk electrode (RDE) as an experimental tool (56.2A). The famous Levich equation describes the electric current as a function of the rotation speed of an RDE:

$$I = (0.62)\ n\,F\,A\,D^{2/3}\,\omega^{1/2}\,v^{-1/6}\,C$$

where I is the measured current (A); n is the number of moles of electrons transferred in the half reaction (number); F is the Faraday constant (C/mol); A is the electrode area (cm^2); D is the diffusion coefficient (cm^2/s); ω is the angular rotation speed of the RDE (rad/s); v is the kinematic viscosity (cm^2/s); C is the concentration of redox species (mol/cm^3) in this form of the equation the constant with a value of 0.62 has units of $rad^{-1/2}$.

The Levich equation can be simplified by defining a Levich constant B:

$$I = B\,\omega^{1/2};\ \text{where } B = (0.62)\ n\,F\,A\,D^{2/3}\,v^{-1/6}\,C$$

According to this equation, there is a linear dependence of the current (I) on the square root of the RDE rotation speed ($\omega^{1/2}$). This linearity can be preserved when the electrochemical process is mass-transport limited (not being limited by the interfacial electron transfer rate). When the rotation speed becomes very fast, the process might be already limited by the electron transfer rate, thus demonstrating deviation from the linear dependence predicted by the Levich equation (56.2B).

Levich fundamental contributions to theoretical physics and physical chemistry have been recognized worldwide after the publication of the English translation of his monograph *Physicochemical Hydrodynamics* in 1962 (56.3). His research activities also included gas-phase collision reactions and the quantum mechanics of electron transfer.

Levich was one of the first Soviet scientists who succeeded in emigrating from the USSR in 1978, first to Israel and then to the USA. (Note that in the Soviet time, the immigration was almost impossible). Then he established the Institute of Applied Chemical Physics that is named now Benjamin Levich Institute for Physico-Chemical Hydrodynamics.

56.1. Benjamin Levich (Veniamin Grigorievich Levich)

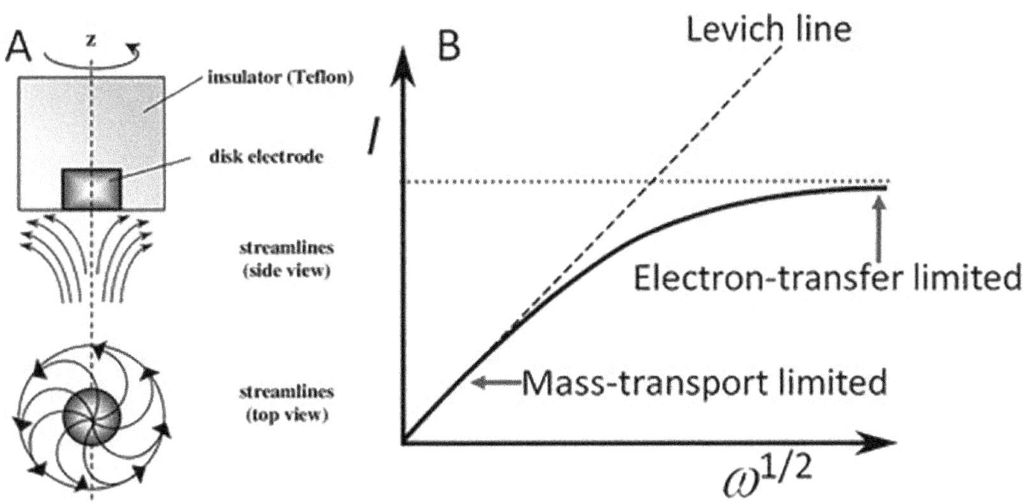

56.2. (A) Mass transport at an RDE. (B) Levich plot corresponds to the electrochemical process at an RDE with different rotation speeds applied. The plot shows deviation from the linear function when the electrochemical process starts to be limited by the interfacial charge transfer at the high electrode rotation speed.

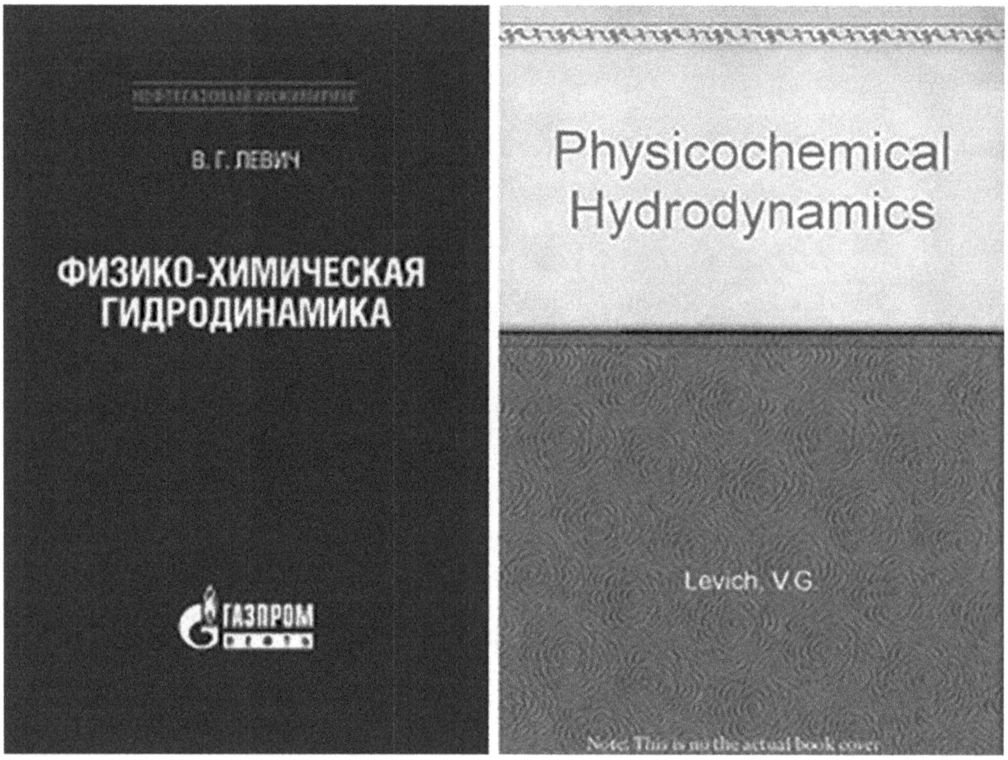

56.3. Famous Levich book, *Physical Hydrodynamics* (original Russian book, 1959, second edition [*left*]; English translation, 1962 [*right*]).

57

LELAND CHARLES CLARK

Born: December 4, 1918, Rochester, New York, USA
Died: September 25, 2005, Cincinnati, Ohio, USA

Leland Charles Clark (57.1) was an American biochemist who is famous mostly for inventing an electrochemical device (Clark electrode) for measuring O_2 concentration in liquids, which was later converted by him to the very first electrochemical biosensor upon placing glucose oxidase (GOx) enzyme on a membrane coating the electrode compartment. Therefore, Clark is recognized as a father of biosensors. The modern-day glucose biosensors used daily by millions of diabetics are conceptually based on his research (still technically being very different).

The Clark electrode is an electrode that measures oxygen concentration in a liquid using a platinum (Pt) working electrode according to the net reaction:

$$O_2 + 4\,e^- + 4\,H^+ \rightarrow 2\,H_2O$$

The Pt electrode is negatively polarized versus Ag/AgCl reference electrode to allow the concentration-limited cathodic (reductive) current. Both electrodes (Pt and Ag/AgCl) are placed into a sensor compartment filled with an electrolyte solution (usually saturated aqueous KCl solution). The compartment is closed with an O_2-permeable membrane (using a dialysis membrane), which allows only O_2 penetration into the measuring compartment and restricts other redox species to get inside the O_2 sensor. This membrane provides the "current response only" dependent on the O_2 concentration in the analyzed liquid.

Despite the simplicity of the invented Clark electrode, the importance of this invention was immediately recognized, and when the paper was published (in 1956), it became one of the most cited papers in the life sciences. The electrode, in its original form, has been in use in a broad range of applications requiring precision O_2 measurements, including water quality assessment, many medical and pharmacological situations, and food production. Clark's electrode allowed oxygen to be monitored actually during surgery, an invention that has saved millions of lives over the past half century.

It is interesting to note that exactly the same device can be used for measuring H_2 concentration instead of O_2 when the Pt electrode is positively polarized (opposite to the case when O_2 is measured), resulting in the H_2 electrochemical oxidation.

The first glucose biosensor (57.2) invented by Clark was a simple device where the O_2-permeable membrane was functionalized with glucose oxidase (GOx) enzyme consuming O_2 in the presence of glucose.

$$\text{glucose} + O_2 \text{ (GOx)} \rightarrow \text{gluconic acid} + H_2O_2$$

The O_2 concentration was inversely dependent on the glucose concentration; thus, the current response was decreased proportionally to the glucose concentration. The membrane performed two functions: it provided support for the immobilized enzyme, and it restricted other redox species to penetrate inside the biosensor's compartment where both electrodes (Pt and Ag/AgCl) are located. The biosensor device provided specific response to glucose concentration regardless of the complex biochemical composition of the analyzed biological liquid (e.g., blood).

Clark performed pioneering research on artificial heart-lung machines in the 1940s–1950s (57.3A) and obtained over twenty-five patents. Particularly, he developed a fluorocarbon-based liquid that could be breathed successfully by mice in place of air. His lifelong goal of developing artificial blood remained unfulfilled at the time of his death, being still not fully achieved now. He is the inventor of oxycyte (based on perfluoro *tert*-butylcyclohexane) (57.3B), a third-generation perfluorocarbon (PFC) therapeutic oxygen carrier designed to enhance oxygen delivery to damaged tissues.

57.1. Portrait of Dr. Leland C. Clark Jr. on or about the date of being awarded the 2005 Fritz J. and Dolores H. Russ Prize, "for bioengineering membrane-based sensors in medical, food, and environmental applications."

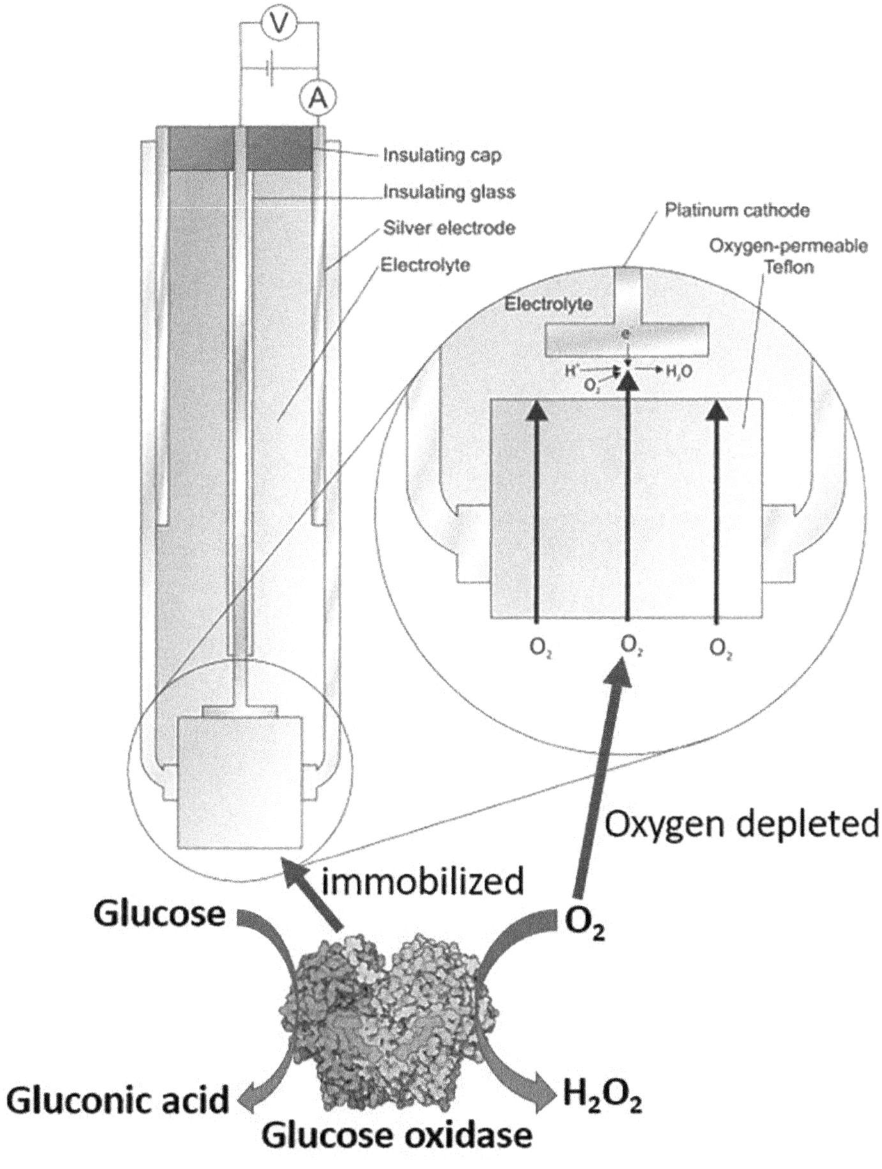

57.2. Schematic presentation of the first glucose biosensor based on the O_2 Clark electrode with the glucose oxidase enzyme immobilized at the O_2-permeable membrane. Notably, immobilization of other oxidase enzymes (e.g., lactate oxidase) consuming oxygen in the enzyme-catalyzed reaction can result in other biosensors operating according to the same concept.

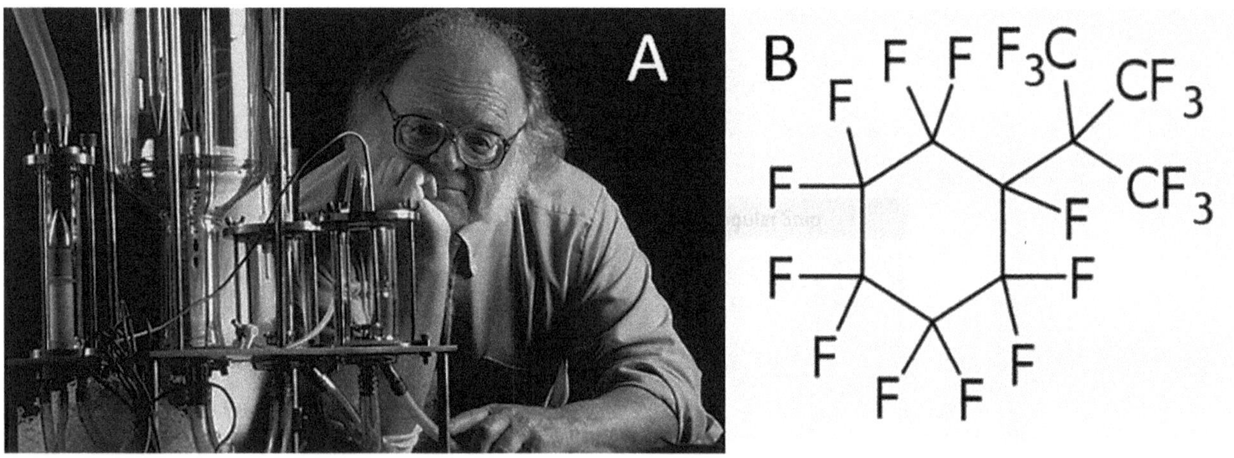

57.3. (A) Clark with a heart-lung machine. (B) Perfluoro *tert*-butylcyclohexane—artificial O_2 carrier claimed to carry oxygen with up to five times the efficiency of hemoglobin when used as an intravenous emulsion.

58

HEINZ GERISCHER

Born: March 31, 1919, Wittenberg, Germany
Died: September 14, 1994, Berlin, Germany

Heinz Gerischer (58.1) was a German electrochemist concentrated in his studies on electrochemistry of semiconductor electrodes with applications in photoelectrochemistry. Gerischer made important contributions not only to electrochemistry but also to physical chemistry and materials chemistry.

From 1949 to 1955, Gerischer was interested in electrode kinetics and developed instruments and techniques for their study, including the electronic potentiostat. (Note that time potentiostats were not broadly available.) He used these instruments for monitoring fast electrochemical processes by double potential step and alternative current (AC) modulation methods. This work lay the foundation for a mechanistic interpretation of electrode reactions and impacted vastly on our present understanding of electrode kinetics.

In 1954–1961, Gerischer began his work with semiconductor electrochemistry, particularly concentrating on the electrochemistry of n- and p-type germanium. Gerischer recognized the enormous implications of semiconductor electrochemistry, both for theoretical studies of charge transfer and for potential applications in photochemistry and photovoltaic devices. He studied Faradaic reactions of electrons and holes (1959) and the theory of electron tunneling at semiconductor electrolyte interfaces.

In 1950s–1960s, Gerischer continued his studies of metal electrodes with papers on the electronic potentiostat (1957), stress corrosion (1957), hydrogen evolution, and hydrogen adatom formation (1957), on the monitoring of fast electrode processes (1960), and on the reaction kinetics of water dissociation as probed by the microwave pulse method (1961).

In 1964–1968, Gerischer, with his research team, studied photoelectrochemistry and photosensitization on electrode materials, such as ZnO, CdS, GaAs, silver halides, anthracene, and perylene. These studies made his group the worldwide-leading experts in semiconductor photoelectrochemistry.

Gerischer published more than three hundred papers in the areas of electrochemical kinetics, fast reactions in solution and at electrodes, electrocatalysis, metal deposition and corrosion, semiconductor electrochemistry, and photoelectrochemistry. He also served as the editor of *Advances in Electrochemistry and Electrochemical Engineering* (with C. W. Tobias), 1977–1984; *Advances in Electrochemical Science and Engineering* (with C. W. Tobias), 1990–1994; and *Light-Induced Charge Separation in Biology and Chemistry*, 1979.

58.1. Heinz Gerischer

59

VLADIMIR SERGEEVICH BAGOTSKY

Born: January 22, 1920, Bern, Switzerland
Died: November 12, 2012, Colorado, USA

Vladimir Sergeevich Bagotsky (59.1) was a Russian/Soviet (with Jewish ethnicity) scientist who greatly contributed to many different areas of electrochemistry, particularly to the design of novel electrical batteries. Despite the fact that he was born and spent his childhood in Switzerland, his lifelong scientific career was in the USSR (presently Russia). Then for retirement, Bagotsky immigrated to the USA.

Bagotsky started his work on electrochemistry being a student. One of his early research topics was study of electrochemical hydrogen evolution, which is a process with a rather complicated mechanism and kinetics. Alexander Naumovich Frumkin (famous Russian electrochemist, 1895–1976) was his first scientific advisor. Then Bagotsky successfully solved a theoretical problem of concentration polarization in irreversible electrode processes. Certain mathematical solutions were first derived by Naum Natanovich Meiman (Russian/Soviet mathematician, 1912–2001). Then Bagotsky successfully applied the results to the dropping mercury electrode used in polarography. The analysis of polarographic data in the mixed kinetic regime is presently known as the Meiman-Bagotsky technique.

Bagotsky studied the operation of porous electrodes and invented a method to measure their porosity. He designed and optimized gas diffusion (breathing) electrodes. Bagotsky studied various electrocatalysts, including those based on Pt microparticles deposited on catalytically inactive pyrolytic graphite electrodes. Notably, he found that the catalytic activity of the Pt species was inversely dependent on the size of the microspecies. In other words, the smaller Pt particles demonstrated, the higher catalytic activity.

Bagotsky, with colleagues, studied the anodic oxidation of a large number of low molecular weight organic substances, such as formic acid, oxalic acid, formaldehyde, glycerol, butanol, glucose, ethylene glycol, etc. Later, this study was extended to methanol adsorption and oxidation on a platinum electrode, which demonstrated catalytic activity for the anodic oxidation process. To generalize the mechanism of the anodic catalysis other noble metals—palladium, iridium, rhodium, and ruthenium electrodes—were studied. Based on the extensive experimental data, Bagotsky formulated a general mechanism for the electrochemical oxidation of organic substances.

The electrochemical studies performed by Bagotsky were concentrated on applied topics, particularly related to electrochemical batteries. He designed batteries for submarines, airplanes, and spacecrafts. These electrical power devices included silver-zinc, water-activated, and thermal-reserve batteries. Particularly, Ag-Zn batteries made by Bagotsky were used in the first Soviet satellite (*Sputnik*) in 1957 (59.2). Similar batteries were later used in other Russian space satellites, including the first crewed spaceflight with cosmonaut (*astronaut*) Yuri Gagarin in 1961. Although much of his applied work was carried out in secrecy (for obvious reasons), Bagotsky nevertheless managed to publish a small amount of fundamental research concerning the electrochemical power sources and other topics (59.3). The famous textbook coauthored by Bagotsky is still important: Frumkin AN, Bagotsky V. S., Iofa Z. A., Kabanov B. N., *Kinetics of Electrode Processes,* 1967 (the original version in Russian was published in 1952) (59.3).

59.1. Vladimir Sergeevich Bagotsky

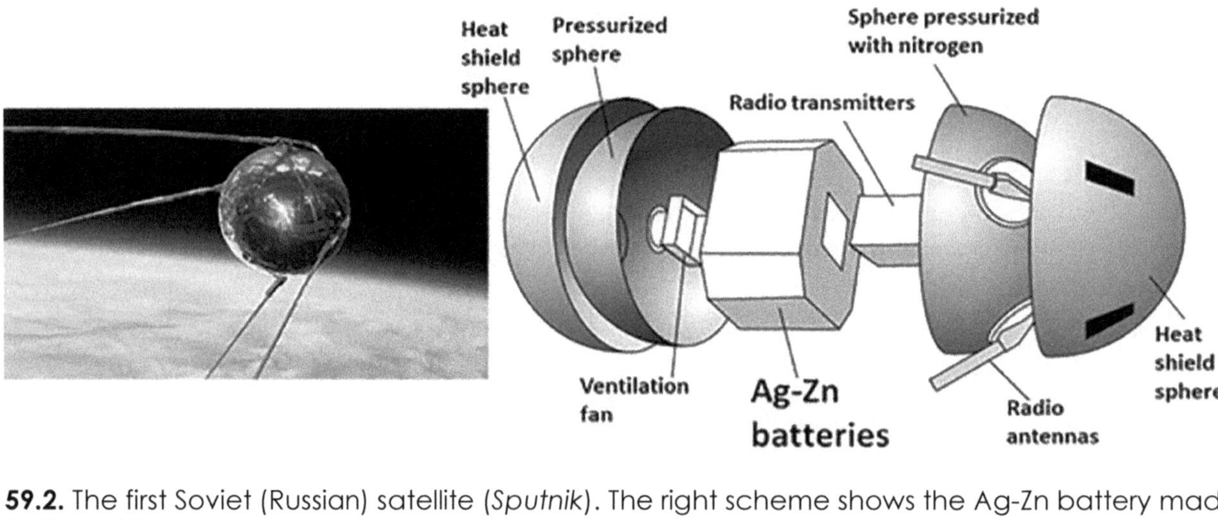

59.2. The first Soviet (Russian) satellite (*Sputnik*). The right scheme shows the Ag-Zn battery made by Bagotsky.

59.3. A few books coauthored by Bagotsky (the first from the right is the famous textbook: *Kinetics of Electrode Processes*—an original Russian version, 1952).

60

JAROSLAV KOUTECKÝ

Born: October 14, 1922, Kroměříž, Czechoslovakia (presently Czech Republic)
Died: August 10, 2005, Berlin, Germany

Jaroslav Koutecký (60.1) was a Czech physical chemist and electrochemist (worked also in the USA and Germany). His broad scientific interests included theoretical electrochemistry, quantum chemistry, quantum theory of surface phenomena, photochemistry, and theory of clusters, among many other topics. In theoretical electrochemistry, Koutecký solved differential equations describing spherical diffusion (relevant to a diffusion process at a mercury dropping or hanging drop mercury electrodes), slow electrode reactions, kinetic currents, catalytic currents, and nonlinear electrochemical reactions. Koutecký's wide-ranging interests have included theoretical treatments of both metal and semiconductor surfaces. His particularly important contribution was in the analysis of electrochemical reactions proceeding at a rotating disk electrode (RDE).

In 1947, Koutecký derived the first theoretical solution of the differential equations for the polarographic currents measured on a dropping mercury electrode (DME) by ignoring the hydrodynamic effects at the DME. Unfortunately, the results were in rather poor agreement with experimental data. Then in 1951, Koutecky realized that the hydrodynamics must be included in the differential equations together with the diffusion and chemical reactions. From that time, Koutecký always included mass transport contribution to electrochemical processes.

The most important equations describing electrical current dependent on the kinetic activity and mass transport of redox species were formulated by Koutecký in cooperation with Veniamin Grigorievich Levich (Russian electrochemist with Jewish ethnicity, worked in Russia and then in the USA, 1917–1987) and named as Koutecký-Levich equations.

The Koutecký-Levich equation (60.2A), formulated in the most general way, has similarity with the conductance of a parallel electrical circuits.

$$\frac{1}{i_m} = \frac{1}{i_K} + \frac{1}{i_{MT}}$$

where i_m is the measured current (A), i_K is the kinetic current limited by the electron transfer rates (A), and i_{MT} is the mass transport current (A); A is an ampere unit. The kinetic current (i_K) can be modeled by Butler-Volmer equation. Note that this current is potential dependent. On the other hand, the mass transport current (i_{MT}) is potential independent and depends on the electrode/cell geometry and stirring. In the very special case of a rotating disk electrode (RDE), the i_{MT} can be modeled using Levich equation. Using Levich equation to express the mass transport current as the function of the RDE rotating speed, Koutecký-Levich equation can be reformulated:

$$\frac{1}{i_m} = \frac{1}{i_K} + \frac{1}{B_L \omega^{0.5}}$$

where: B_L is the Levich constant and ω is the angular rotation speed of the electrode (rad/s). Note the square root ($\omega^{0.5}$) dependence of the current on the rotation speed originated from Levich equation.

Following the current dependence on the rotation speed of a RDE corresponding to Levich theory, the current is measured at different rotation speeds. The kinetic current can be extracted from Koutecký-Levich plot (60.2B). The linearized data allows extrapolation to the ordinate, where the reaction is not mass transport limited (assuming infinitely high rotation speed). Therefore, Koutecký-Levich analysis allows to determine the kinetic constants of the electrochemical reaction, such as the rate constant, k^0, and the symmetry factor, α.

The approach developed by Koutecký and Levich is fundamentally important for the analysis of electrochemical kinetics.

60.1. Jaroslav Koutecký

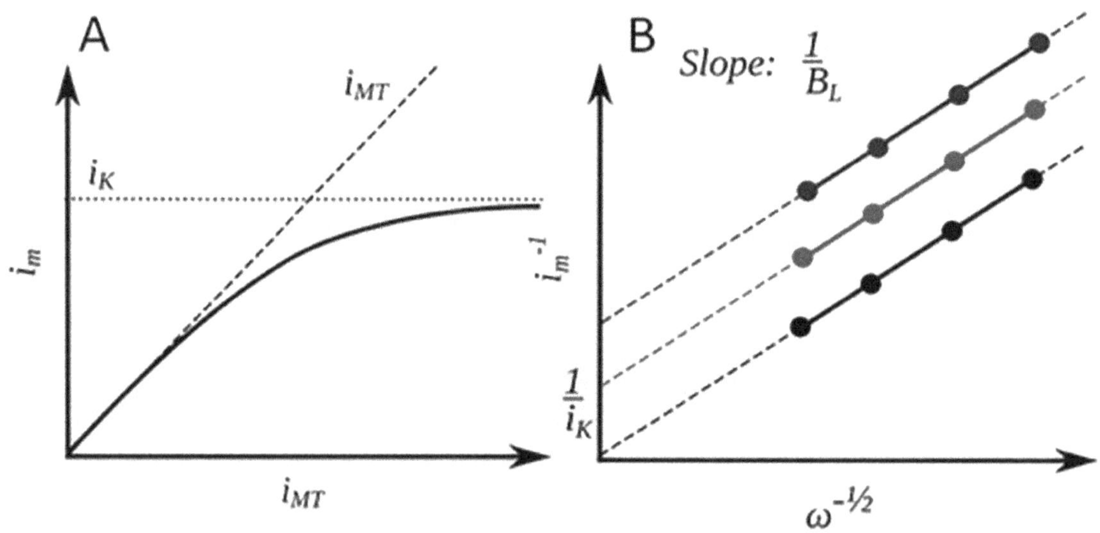

60.2. (A) A visualization of Koutecký-Levich equation. The graph shows the measured current as a function of the mass transport current for the given kinetic current. (B) Koutecký-Levich plot. The inverse measured current is plotted versus the inverse square root of the rotation speed of a rotating disk electrode (RDE).

61

JIŘI KORYTA

Born: June 24, 1922, Pisek, Czechoslovakia (presently Czech Republic)
Died: May 30, 1994, Prague, Czech Republic

Jiři Koryta (61.1) was a Czech electrochemist particularly known for his famous and unique research on electrochemical effects at an interface of two immiscible liquids.

Koryta studied the electrochemistry of complex compounds and the adsorption of electrochemically inactive species at electrode surfaces. While these studies were done with very high experimental quality, they were still quite common among many electrochemists. More importantly, he performed research on several rarely studied systems, including mercury jet (streaming) electrode (61.2). This electrode is not commonly used in electrochemistry because of the huge amount of mercury involved in the experiments, thus requiring special precautions to exclude poisoning with mercury vapor.

Later, Koryta concentrated on the research of electrochemical phenomena at an interface between two immiscible liquids, including, for example, a water/nitrobenzene interface (61.3). In these studies, he observed ion (e.g., alkali metal and ammonium ions) transfer across the liquid/liquid interface, especially facilitated by crown esters producing complexes with the cationic species. These unique studies were the most important contribution made by Koryta to electrochemistry.

The interface between two immiscible electrolyte solutions plays vital roles in various fields, such as neuroscience, environmental science, analytical chemistry, separation, catalysis, and nanoparticle synthesis. This research area emerged as a unique approach in electrochemical analysis due to its sensitivity to both redox-active and redox-inactive analytes. Following innovative research performed by Koryta, neurotransmitters, metal ions, drug molecules, and other complex molecules such as proteins have been detected.

Koryta was also interested in theoretical electrochemistry, bioelectrochemistry, and the history of natural science.

61.1. Jiři Koryta

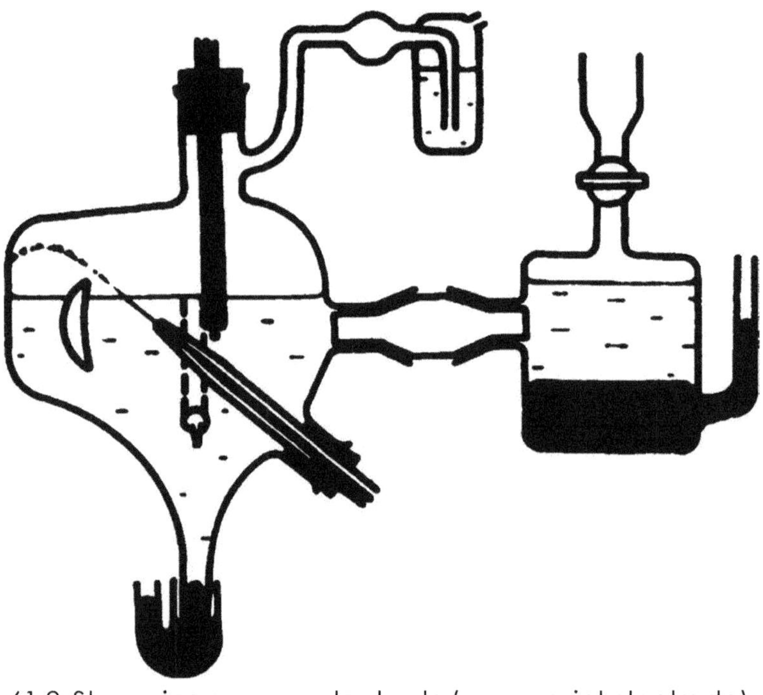

61.2. Streaming mercury electrode (mercury jet electrode). In the left part, a streaming mercury electrode is shown together with a dropping mercury electrode, and in the right part, a large mercury pool is shown, which served as a counter electrode having a constant potential.

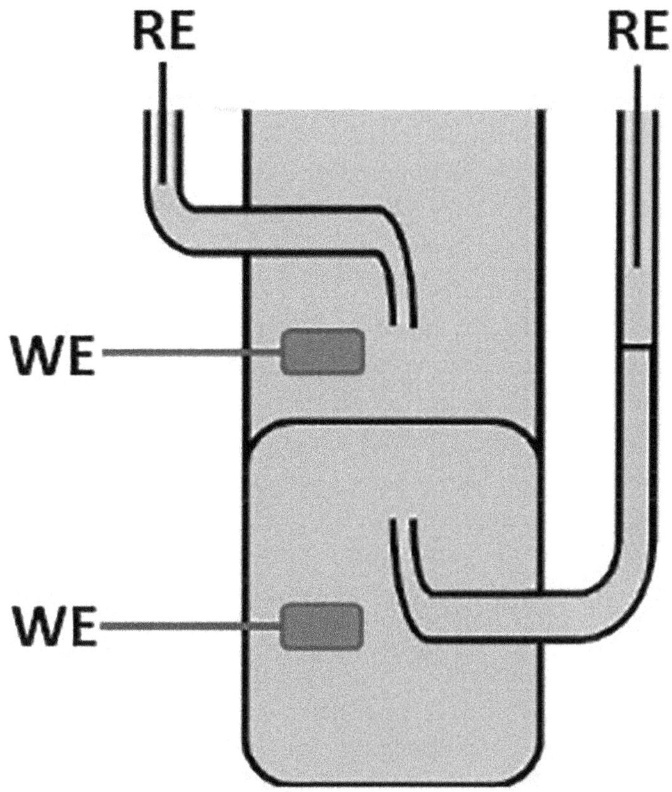

61.3. A schematic of the interface between two immiscible electrolyte solutions, including an organic and an aqueous phase, set up for electrochemical measurements. Each phase contains a working and reference electrode (WE and RE, respectively), which allow the application of potential difference (voltage) between the two liquid phases. The current measured between two WEs reflected the ion transfer from one liquid phase to another one.

62

HANS WENKING

Born: August 18, 1923
Died: June 19, 2007

Hans Wenking (62.1) was a German electrical engineer, physicist, and inventor who devoted his lifetime to the development of electronic equipment for chemistry and physics, particularly constructing the first potentiostats, which became major parts of modern electrochemical instruments. He was the first who described the basic principles of potentiostats.

In 1952, Hans Wenking constructed an electronic amplifier for controlling an oscilloscope using a mirror galvanometer with the signals recorded on photographic paper. This amplifier was later used as a core of a new potentiostat with only a power supplier added. This potentiostat, being an important instrument for electrochemical investigations, was designed by Wenking during his work at the Max Plank Institute in Göttingen, Germany.

At that time, Hans Wenking was working in a group with Prof. Karl Friedrich Bonhoeffer (German chemist, 1899–1957) where he was appointed to develop a potentiostat that was needed for electrochemical experiments, particularly for studying corrosion. Until 1957, Wenking's potentiostat was manufactured only for the internal use of the Max Planck Institute in Göttingen. Later, Hans Wenking, together with Gerhard Bank, established Elektronische Werkstatt Göttingen to commercialize the potentiostats. From 1959, the company operated under the name Gerhard Bank Electronik. Wenking designed the instruments as a freelance, but the brand Wenking potentiostat (62.2) soon became a famous trademark. The consequence of the potentiostat development was a rush in the development of electrochemical science. The potentiostats became standard instruments for most electrochemical investigations, particularly for electroanalytical measurements.

Independent of Wenking's work, similar instruments were designed by other companies. Tacussel was one of those companies that came to a similar design as Wenking's potentiostats and started manufacturing potentiostats in France. In the USA, Wenking's potentiostats were leading on the market and became standard instruments in electrochemical labs.

Wenking has never published his results in scientific papers. On the other hand, Wenking never concealed the technical details of his instruments. The circuits and layouts designed by him were included in the operation manuals for the instruments, and even in some manuals, a detailed theoretical treatise was given. Wenking's instruments were always state-of-the-art, and electrochemists of the 1950s–1970s used them for many different applications.

Moving from two-electrode to three-electrode configuration—Potentiostats for electroanalysis and constant potential electrolysis

In the 1920s–1950s, many polarographic and later voltammetric (e.g., cyclic voltammetry) measurements were performed using a two-electrode configuration composed of a dropping mercury electrode or any other small working electrode and a counter electrode, also serving as a reference. The two-electrode configuration provided reasonable quality of the electrochemical measurements as long as the working electrode was small, the potential sweep was not fast, and the background solution contained a high concentration of an electrolyte in an aqueous solution, providing a small current over small resistance in a liquid phase. In other words, the potential drop in the electrolyte solution was not significant to affect the potential measurement precision, so the applied voltage was close to the actual potential of the working electrode. However, when the analytical measurements started to be performed in nonaqueous solvents with much higher resistance and/or the potential scan rates were much faster (particularly in cyclic voltammetry), resulting in larger currents, the potential drop in the liquid phase became significant, thus requiring its compensation.

At this time, the standard electrochemical/electroanalytical instruments included a three-electrode configuration with a reference electrode added. The three electrodes—working, counter, and reference—have operated with a potentiostat, having more sophisticated electronic circuitry than previously used two-electrode polarographs. While sometimes the reference electrode is not really needed or difficult to be used—particularly in micro-sized electrochemical cells or in implantable electrochemical devices operating in biological fluids—modern electroanalytical devices have always the circuit, allowing their operation with three electrodes.

The transition from electrochemical instruments operating with two electrodes to those operating with three electrodes, functioning as potentiostats, occurred in approximately the 1950s when numerous potentiostats became commercially available (62.3 and 62.4). Many of them were based on the fundamental design of Hans Wenking. Some of these devices were used for controlled potential electrolyzers rather than electroanalytical applications. Notably, the electrolysis was conducted with large-area electrodes, which operated with large currents and required very powerful electrical devices. It should be also mentioned that many polarographs produced in 1950 and later have already included potentiostat electrical circuits.

A multipurpose electroanalytical instrument constructed by Lingane (62.4a) operated with various functions, including the potentiostat function that maintained a constant potential at the working electrode during electrolysis. This mode of the instrument operation was used in electrogravimetric analysis of metals, electrolytic separations of metals prior to final determinations by other methods, coulometric analysis, and electrolytic preparations using the controlled potential technique. Electrolysis with either constant total applied voltage or constant current (i.e., operating in potentiostatic or galvanostatic regimes) was also possible.

Starting from 1950s, all polarographs and later electrochemical analyzers were always equipped with a potentiostat function (62.5). The author of this book used a polarograph with a potentiostat function in his first experiments during his PhD study (62.5C).

62.1. Hans Wenking

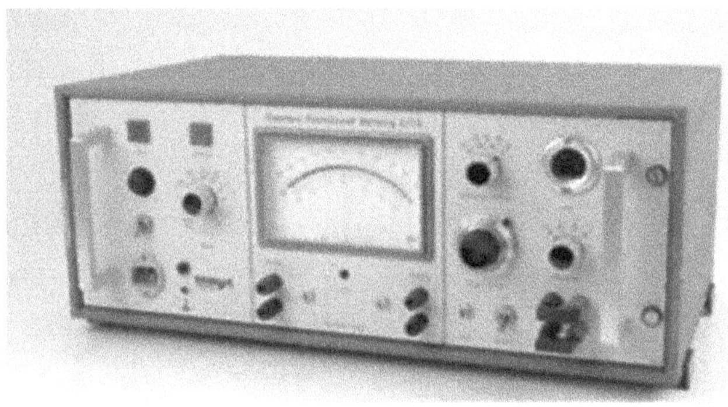

62.2. Bank Elektrotechnik Wenking ST72 Standard Potentiostat ST 72

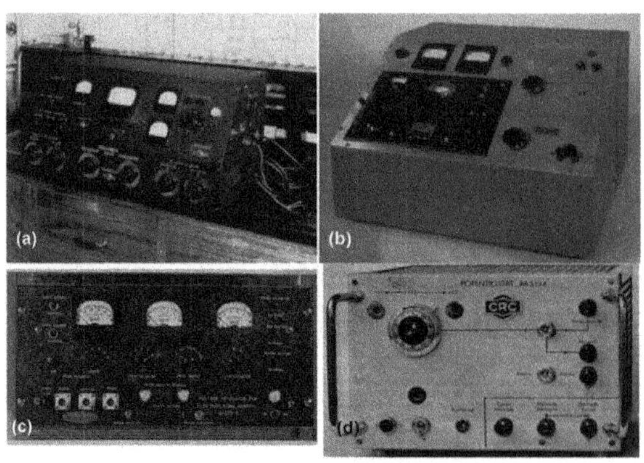

62.3. (a) Caldwell, Parker, and Diehl potentiostat. (b) Penther-Pompeo potentiostat. (c) Lamphere-Rogers potentiostat. (d) Potentiostat PAS 134.

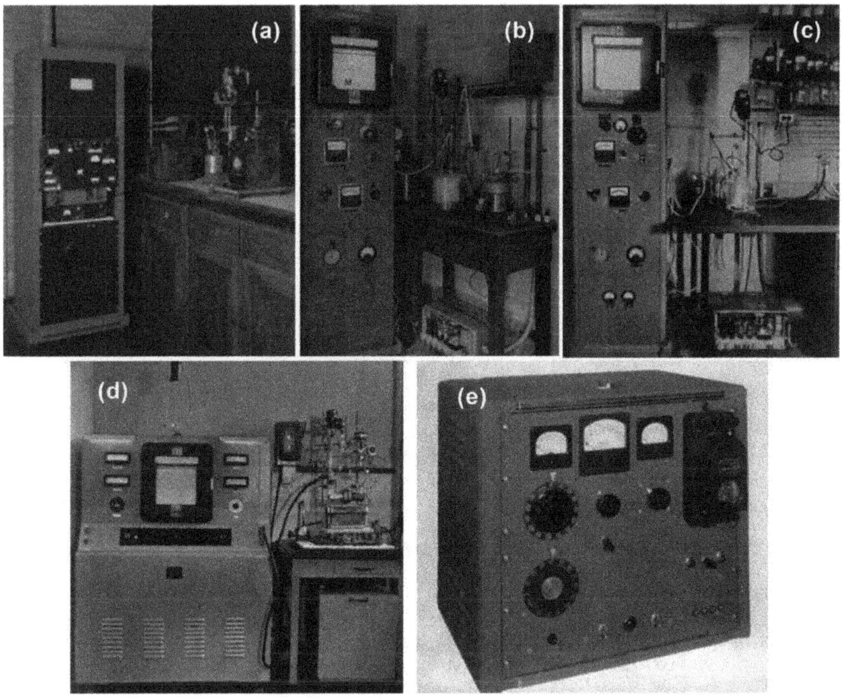

62.4. (a) Lingane multipurpose electrochemical servo instrument. (b) Allen potentiostat. (c) Redoxotrol Model II. (d) Redoxotrol Model III. (e) Lingane-Jones potentiostat.

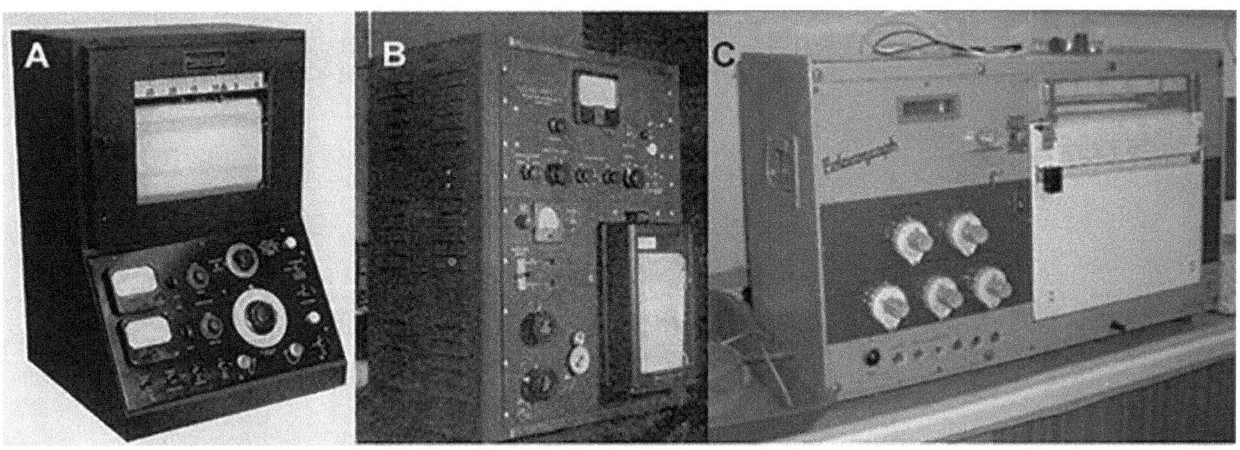

62.5. (A) Visible recording polarograph (Model XXI) manufactured by E. H. Sargent and Co., Chicago, USA. (B) Polaro-Analyzer designed by Schulman-Battey-Jelatis and manufactured by the Rutherford Instrument Co., Alexandria, Virginia, USA. (C) Polarograph Model OH-102 produced by Radelkis, Hungary.

63

BERNHARDT PATRICK JOHN O'MARA BOCKRIS

Born: January 5, 1923, Johannesburg, South Africa
Died: July 7, 2013, Gainesville, Florida, USA

Bernhardt Patrick John O'Mara Bockris (63.1) was a South African-English-American scientist with very broad interests, particularly related to different aspects of electrochemistry, including physical/theoretical electrochemistry, photoelectrochemistry, and bioelectrochemistry. He also contributed to environmental chemistry.

Bockris's best-known work was the theoretical modeling of the electrode-electrolyte interface to include the dipole moment of the solvent molecules (particularly dipoles of water molecules) (63.2). In 1963, Bockris published in the Proceedings of the Royal Society (London) his most important paper on this topic, *On the Structure of Charged Interfaces*. This work was coauthored with Michael A. V. Devanathan and Klaus Müller. The paper is widely referred to as the BDM paper (Bockris-Devanathan-Müller).

Experimental and theoretical work on electrochemical kinetics was Bockris major contribution to electrochemistry. He also performed studies of high-temperature chemistry (particularly the electrochemical behavior of silicates at high temperature, 2,000°C); analysis of factors determining electrolyte solubility; and hydrogen/metal interactions. Bockris considered the hydrogen evolution reaction as the prototype for all other electrochemical reactions. Bockris performed a series of classic experiments with Kohei Uosaki on the photoelectrochemical generation of hydrogen, which he considered as the bases of future hydrogen energetics.

For promotion of education in electrochemistry, he wrote (coauthored by Amulya K. N. Reddy) the famous two-volume textbook *Modern Electrochemistry*. Bockris was at the center of physical chemistry for more than fifty years and was among the top ten most cited physical chemists.

63.1. John Bockris

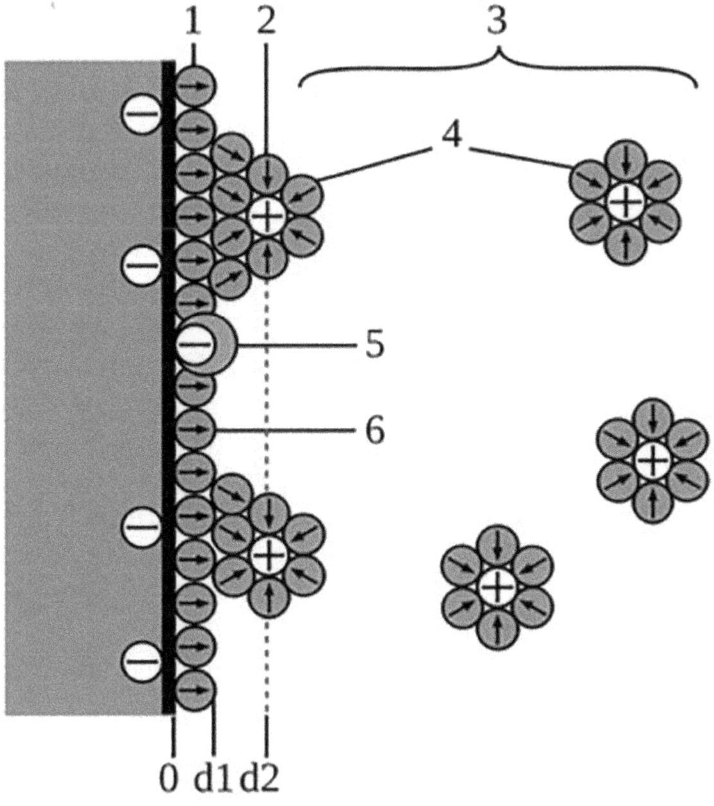

63.2. Schematic representation of a double layer on an electrode (BDM) model. (1) Inner Helmholtz plane (IHP); (2) outer Helmholtz plane (OHP); (3) diffuse layer; (4) solvated ions (cations); (5) specifically adsorbed ions (redox ions, which contribute to the pseudocapacitance); (6) molecules of the electrolyte solvent. The d_1 and d_2 are thicknesses of the IHP and OHP layers, respectively.

64

PETR ZUMAN

Born: January 13, 1926, Prague, Czechoslovakia (presently Czech Republic)
Died: June 24, 2021, Potsdam, New York, USA

Petr Zuman (64.1) was a Czech-origin, American electrochemist mostly remembered for his work on polarography of organic substances, particularly analyzing mechanisms of organic reactions.

Zuman was introduced to the polarography method while he was working with Jaroslav Heyrovský (Czech scientist who invented polarography and received Nobel Prize, 1890–1967). Zuman applied polarography (electroanalytical method with mercury droplets used as a working electrode) and voltammetry (using solid metal electrodes) for the analysis of mechanisms of some organic reactions. For example, he studied acid-base, hydration-dehydration, and tautomeric equilibria involving 1,3,5- and 1,2,4-triazines, selenous acid, and mitomycin C. He also analyzed reactions of bile acids, cholesterol, and other sterols in strongly acidic media, as well as the additions of nucleophiles, such as glutathione, to nitrosobenzene. Many of his studies included biological molecules. These studies were important for various bioanalytical applications and allowed better understanding of the biological activity of the studied biosubstances.

Zuman investigated structure-reactivity relationships in many organic compounds. For example, he studied electron transfer effects through an N-N single bond, ring formation of some 2-amino-1,4-benzoquinones, and interactions between two reducible groups in a molecule.

Zuman published over five hundred research and review papers as well as fifteen textbooks and reference books (64.2). One of them was his influential monograph, *Substituent Effects in Organic Polarography* (Plenum Press, New York, 1967). In this book, he provided strong links between modern electrochemical methods and physical organic chemistry. He particularly formulated linear free energy relationships to explain substituent effects on redox properties of organic molecules. Petr was the last person in the scientific generation who used polarography as a tool for electroanalytical applications.

64.1. Petr Zuman

64.2. Books on polarography written by Zuman

65

ALAN MACDIARMID

Born: April 14, 1927, Masterton, New Zealand
Died: February 7, 2007, Drexel Hill, Pennsylvania, USA

Alan MacDiarmid (65.1) was a New Zealand–born American chemist and one of three recipients of the Nobel Prize for Chemistry in 2000, which was given "for the discovery and development of conductive polymers." The discovered conducting polymers opened a new era in electrochemistry, particularly being used in novel modified electrodes important for many applications.

MacDiarmid performed lifelong research concentrated on polymeric materials having novel features of electrical conductivity. In this work, he collaborated with the Japanese chemist Hideki Shirakawa and the American physicist Alan Heeger. The first paper describing novel electro-conducting materials was published in 1977 titled *Synthesis of electrically conducting organic polymers: Halogen derivatives of polyacetylene,* $(CH)_x$. Then in 2000, all three collaborators shared the Nobel Prize in Chemistry for the discovery of conducting polymers.

While electronic (frequently named *metallic*) conductivity is typical for metals and graphite, organic polymers are usually electrical isolators (having no electronic conductivity). It is interesting to note that some experiments performed much earlier in the nineteenth and mid-twentieth centuries have demonstrated redox behavior in some polymers, particularly in polyaniline and polypyrrole. Still, it was far from the demonstration of real electronic conductivity. It was one step left before the conductive polymers were discovered, and this step was done by MacDiarmid, Hideki Shirakawa, and Alan Heeger.

Their work resulted in the synthesis of novel polymeric materials with conjugated chemical bonds supporting the free move of electrons along the polymer chains and providing electronic conductivity similar to that in metals (65.2). Some of these polymers were already known for many decades, but their conductivity was not demonstrated. Notably, the electrical conductivity properties can be fine-tuned using methods of organic synthesis and by advanced dispersion techniques.

The conductive polymers have found numerous applications, particularly including modification of electrodes providing support for (bio)catalysts and other bioelectrochemical applications.

65.1. Alan MacDiarmid, 2005

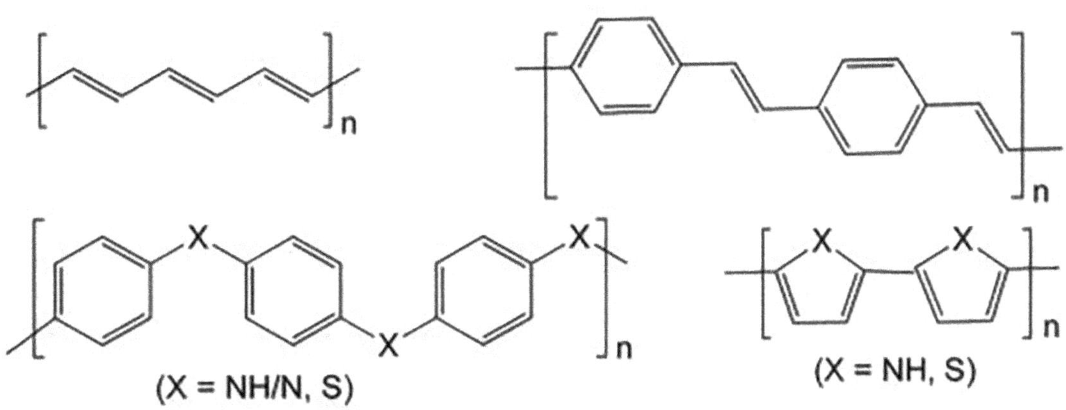

65.2. Chemical structures of some conductive polymers. From top left clockwise, polyacetylene; polyphenylene vinylene; polypyrrole (X = NH) and polythiophene (X = S); and polyaniline (X = NH) and polyphenylene sulfide (X = S).

66

BRIAN EVANS CONWAY

Born: January 26, 1927, Farnborough, United Kingdom
Died: July 9, 2005, Ottawa, Canada

Brian Evans Conway (66.1) was a Canadian chemist/electrochemist of British origin. Based on his pioneered studies, he is considered as a father of supercapacitors.

Conway had a very broad range of scientific interests, and he worked on nearly all aspects of electrochemistry, including the electrified interfaces, ion solvation, adsorption processes, electrode kinetics, solvated electrons, oxide film formation, electrocatalysis, rechargeable batteries, electrochemical capacitors, and particularly supercapacitors (the term he coined).

In 1975–1980, Conway performed fundamental research on ruthenium oxide electrochemical capacitors. Later in 1991, Conway formulated the difference between a supercapacitor (note that this term was introduced later) and a battery, highlighting their different mechanisms for electrochemical energy storage. Then in 1999, Conway introduced the term *supercapacitor* to emphasize the capacitance increase due to surface redox reactions with faradaic charge transfer between electrodes and ions. The supercapacitor stored electrical charge partially in the Helmholtz double layer and partially as a result of faradaic reactions with pseudocapacitance charge transfer of electrons and protons between electrodes and electrolytes. The mechanism of formation of the pseudocapacitance is based on interfacial redox reactions with an additional contribution of intercalation and electrosorption (adsorption onto a charged electrode surface). The research performed by Conway, experimentally and theoretically, greatly expanded the knowledge of electrochemical capacitors.

A typical supercapacitor is an electrochemical device (66.2) with high capacity, which is much higher than in solid state capacitors with a dielectric layer separating two electrodes. The dramatic capacitance increase in supercapacitors is obtained due to two mechanisms of the charge collection: a large interface surface (electrodes with a rough surface and high porosity) and Faradaic redox process. Both mechanisms are responsible for the charging/discharging the electrode-electrolyte interface.

A supercapacitor typically can store from ten to a hundred times more electrical energy (per unit volume or mass) than traditional capacitors based on a dielectric separator between the electrodes. Supercapacitors can deliver a big discharge energy very fast and allow many repeating charge-discharge cycles. Therefore, the supercapacitors are used when fast discharge operation delivering large power rapidly is needed.

While Conway's contribution to the supercapacitor construction, operation, and theory was the most important, he was working in his last decades, 1979–2005, on many other subareas of electrochemistry—particularly the temperature dependence of both the Tafel slope and the symmetry factor, electrosorption of ionic species and its influence on surface oxidation, the fluorine and chlorine evolution reactions, ion hydration in the double-layer, single crystal and alloy electrodes, and many more.

66.1. Brian Evans Conway

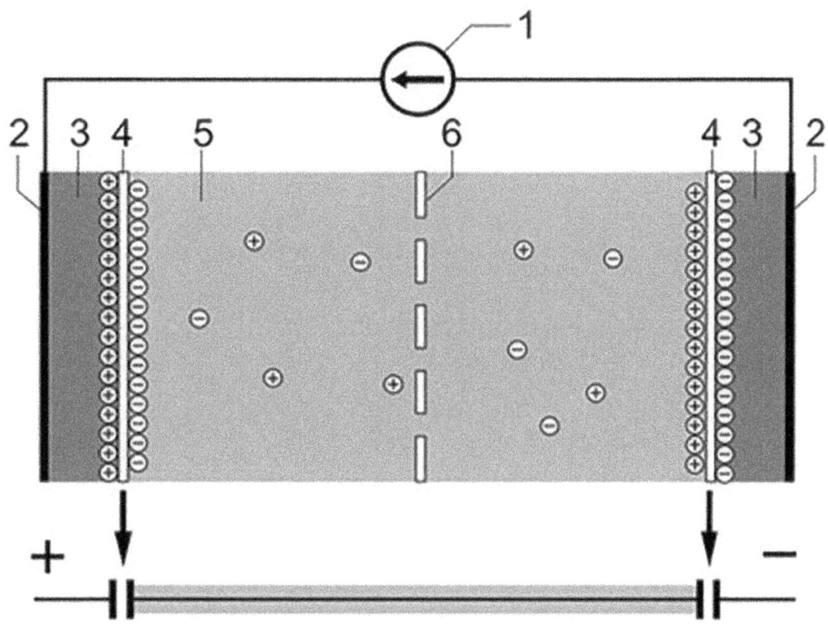

66.2. Schematics of a supercapacitor: (1) power source, (2) power collector, (3) polarized electrodes, (4) Helmholtz double layer, (5) electrolyte containing positive and negative ions, and (6) membrane separator

67

ROBERT ALLEN OSTERYOUNG

Born: January 20, 1927, Cleveland, Ohio, USA
Died: August 10, 2004, Raleigh, North Carolina, USA

Robert Allen Osteryoung (67.1) was an American electrochemist mostly remembered for the use of novel pulse-voltammetry techniques, particularly for the invention of the square wave voltammetry named after him as the Osteryoung square wave voltammetry.

Osteryoung published over two hundred scientific papers on double potential step chronocoulometry, room-temperature molten salt electrochemistry, and square wave voltammetry. Square wave voltammetry (SWV) is an electroanalytical potentiostatic method that offers some advantages to commonly used techniques like cyclic voltammetry. In SWV, the waveform is a series of pulses increasing along a linear baseline, where current is measured in a forward pulse and reverse pulse (67.2). Then the resulting measured response is the current difference between the forward and reverse pulses. The way in which the current is measured at each pulse aids in minimizing the measurement of background (charging) current. The SWV is one of the fastest and most sensitive pulse voltammetry methods. The detection limits can be compared with the most sensitive analytical techniques. The analysis of the characteristic parameters of the SWV enables the evaluation of the kinetics and mechanism of the electrode process under study. Related pulse-type methods include Normal Pulse Voltammetry (NPV) and Differential Pulse Voltammetry (DPV).

Osteryoung also initiated computer-controlled electrochemical measurements, which helped him in developing and optimizing pulse voltammetry. Nowadays, all electrochemical instruments are computer-controlled, and Osteryoung was one of the first scientists who started transition from the analog to digital electrochemical instrumentation.

67.1. Robert Allen Osteryoung

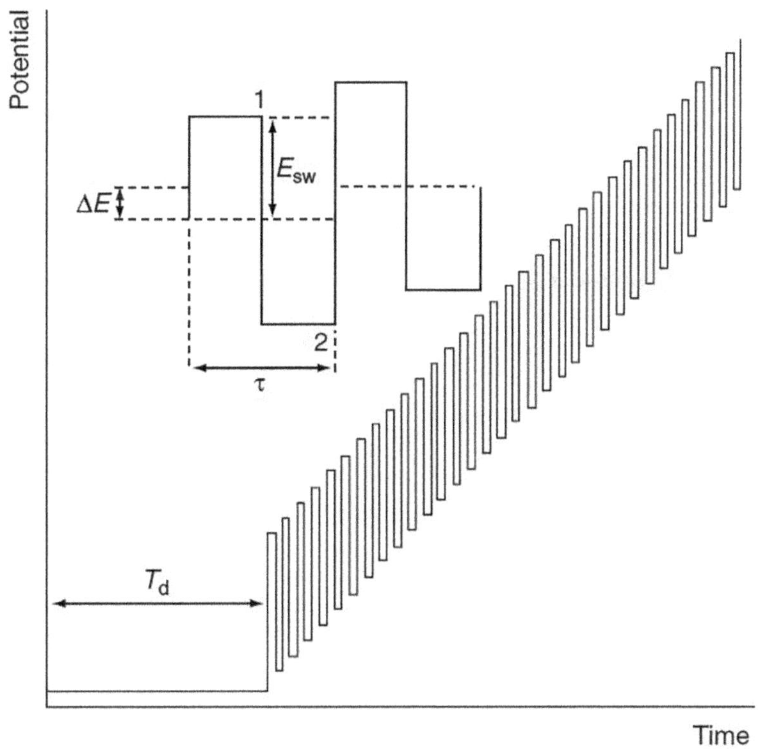

67.2. Square wave waveform showing the amplitude (E_{sw}), step height (ΔE), square wave period (τ), delay time (T_d), and current measurement times 1 and 2.

68

KATSUMI NIKI

Born: March 8, 1931, Kanagawa, Japan
Died: March 4, 2004, Huntington, California, USA

Katsumi Niki (68.1) was a Japanese (working in the USA) electrochemist mostly remembered for his first-time observation of a reversible electrochemical process for a protein.

Niki's most well-known work (1979) is the first discovery of the direct reversible redox reaction of proteins on metal electrodes, particularly demonstrated for cytochrome c3 from sulfate-reducing bacteria (68.2). It should be noted that he used unmodified metal electrodes, working particularly with a hanging mercury drop electrode (HMDE). Note that in this protein, the heme (the redox group in this protein) is located close to the protein backbone periphery, thus allowing electron exchange between the heme and an electrode surface.

68.1. Katsumi Niki

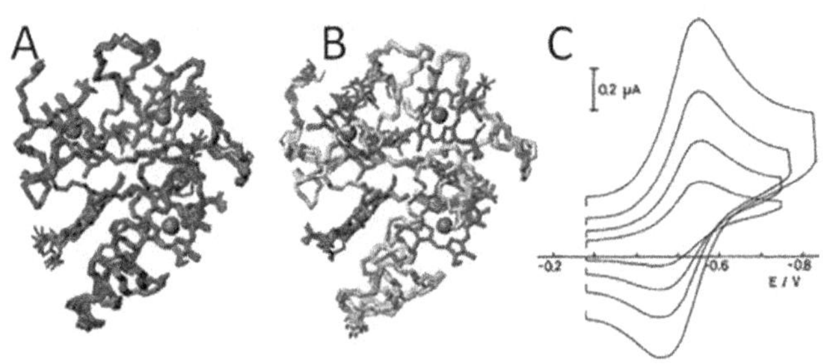

68.2. Solution structures of cytochrome c3 from D. desulfuricans ATCC 27774: (A) oxidized, (B) reduced, (C) cyclic voltammograms of cytochrome c3 on a hanging mercury drop electrode (HMDE) recorded at different potential scan rates.

69

REVAZ "REZO" ROMANOVICH DOGONADZE

Born: November 21, 1931, Tbilisi, Georgia, USSR (presently Georgia)
Died: May 13, 1985, Moscow, USSR

Revaz Dogonadze (69.1) was a Georgian scientist. (Note that in his time, Georgia was a part of the Soviet Union, USSR, and his major work was performed in Moscow.) He greatly contributed to theoretical electrochemistry, particularly introducing quantum mechanics methods to the theoretical treatment of electrochemical reaction processes.

The scientific interests of Dogonadze were closely connected with the theory of kinetic phenomena in condensed media. His pioneering works of 1958–1959 provided the foundations for the modern quantum-mechanical theory of elementary chemical processes in electrolyte solutions. The quantum theory of electrochemical kinetics was further extended in the work performed by him and his colleagues. Dogonadze developed a comprehensive quantum-mechanical theory of the elementary act of electrochemical reactions, considering electron and proton transfer processes at metal and semiconductor electrodes.

Dogonadze was the first scientist who considered a chemical electron transfer process as a quantum-mechanical transition between two distinct electronic states. He introduced nuclear tunneling in the theoretical treatment of the electrochemical processes. Notably, based on his theoretical concept, chemical processes could proceed even at very low temperatures due to the nuclear tunneling mechanism. The electron transfer theory developed by Dogonadze, based on quantum-mechanical transitions, was published in 1959–1961.

This work included the theory of adiabatic and nonadiabatic electron transitions, the role of quantum and classic vibrational degrees of freedom, the role of high-frequency and low-frequency fluctuations of the solvent polarization, the electrode redox reactions, the processes accompanied by light absorption, the low-temperature reactions, the bridge electron transfer reactions, and the electron transfer in biological macromolecules. The criteria for adiabatic and nonadiabatic transfer mechanisms were found, and the method of the multidimensional activated complex was developed for adiabatic reactions.

Most of his results are fundamental and retain importance even now. In particular, he was the first to obtain, by a quantum-mechanical calculation, the expression for the electron transfer probability, which was published in 1959 in his work with Veniamin Grigorievich (Benjamin) Levich (a Russian scientist who later immigrated from the USSR, 1917–1987).

Another fundamental result in his theoretical work was the explanation for the absence of an inversion region in the electrochemical reactions of electron transfer, which was expected from the electron transfer theory developed by Rudolph A. Marcus (Canadian-born American chemist). Dogonadze, with his colleagues, developed the first quantum-mechanical model for the proton transfer. This theory considered the role of the polar medium and a quantum nature of the proton behavior. In particular, the model allowed for the change in the intermolecular distance during the proton tunneling. The joint consideration of the electron and proton transfer steps was the pioneering work that was highly important for understanding electrochemical elementary processes. Dogonadze with colleagues developed the polar-liquid model based on a set of effective oscillators. This model employed the oscillator strength, considering the dielectric properties of the liquid phase.

The scope of Dogonadze scientific interests was very wide. He conducted a number of studies on the theory of low-velocity electrons in disordered systems, theory of solvated electrons, and theory of photochemical processes in solutions. He made an impressive contribution to the theory of elementary biochemical processes. Dogonadze was interested in the nontraditional methods of investigating interfaces, particularly considering the electrodynamics of interfaces and processes of vibrational relaxation.

69.1. Revaz "Rezo" Romanovich Dogonadze

70

JURGEN OTTO BESENHARD

Born: May 15, 1944, Regensburg, Germany
Died: November 3, 2006, Graz, Austria

70.1. Jurgen Otto Besenhard

Jurgen Otto Besenhard (70.1) was a German-Austrian electrochemist who is particularly known for his contribution to the study and construction of lithium batteries, which are highly important for present electronic devices such as cell phones, laptop computers, implantable biomedical devices (e.g., pacemakers), etc.

Besenhard's work was devoted to nonaqueous electrolyte chemistry. While studying electrochemistry in nonaqueous solvents, he was more and more involved in the field of primary and rechargeable lithium batteries. He became one of the fathers of lithium and lithium-ion battery chemistry as we know it today. The scientific background for his research was provided by numerous exploratory research findings on lithium batteries in the late 1960s and early and mid-1970s. While the experimental data were already available, Besenhard brought interpretation and understanding of the complex phenomena involved in the electrochemistry of the Li-batteries. This is especially evident in his early works on the understanding of reversible alkali metal ion intercalation into graphite (in anodes) and reversible alkali metal ion insertion into oxide materials (in cathodes). He summarized and reviewed data on lithium batteries and the preparation of lithium alloys with defined stoichiometry in organic electrolytes at ambient temperature. His pioneering works have been very important for the progress of lithium battery technology, which has become a dominant field in applied electrochemistry since then.

Besenhard expanded his activities to countless topics in the field of applied electrochemistry: electrodeposition, electroless plating, microelectrodes and microelectrode arrays, metallization of plastics, carbon chemistry, intercalation chemistry, composite electrodes, fuel cells, membranes, organic synthesis, electro-fluorination, supercapacitors, coating, aqueous batteries, etc. while continuing his remarkable contributions to the Li battery field. Besenhard was the author or coauthor of more than eight hundred publications and fifty patents and was editor of the *Handbook of Battery Materials*.

71

OTHER SCIENTISTS WHO MADE IMPORTANT CONTRIBUTIONS TO ELECTROCHEMISTRY

In order to mention scientists who made important contributions to electrochemistry over several centuries (mostly in the nineteenth and twentieth centuries), the book should be huge (several volumes and thousands of pages). The scientists, seventy persons, highlighted above were selected based on the author's preferences. Obviously, some names—for example, Volta and Galvani—would be selected anyway even if the book is written by another person. However, many other names might be changed based on the different priorities of a different author. It would be absolutely impossible to mention all names even they are important enough. To do a small step in that direction, twenty additions names are very briefly mentioned in this ending section (71.1). Obviously, this addition cannot cover all important scientists worked in the broad area of electrochemistry.

71.1. (A) Herbert Henry Uhlig (1907–1993) was an American physical chemist who studied corrosion, particularly concentrated on the electrochemical corrosion.
(B) Jean-Michel Savéant (1933–2020) was a French chemist who greatly contributed to the foundation and development of a new discipline—molecular electrochemistry.
(C) Martin Fleischmann (1927–2012) was a British electrochemist focused almost entirely on fundamental electrochemistry. His scientific career was badly affected by misleading reports on "cold fusion."
(D) Ulick Richardson Evans (1889–1980) was a British chemist who played a key role in establishing contemporary understanding of corrosion processes and is often referred to as the father of corrosion science.
(E) Boris Borisovich Damaskin (1932–2019) was a Russian electrochemist who particularly studied adsorption of organic compounds on electrodes.
(F) Allen Joseph Bard (1933–2024) was the most famous American electrochemist—father of modern electrochemistry.
(G) Roger Parsons (1927–2017) was a British electrochemist—one of the leading electrochemists.

(H) Yurii Aleksandrovich Chizmadzhev (1931–2022) was a Russian electrochemist mostly working in the area of bioelectrochemistry and (bio)membranes.
(I) Samuel Ruben (1900–1988) was an American inventor who made lasting contributions to electrochemistry and solid state technology, including the founding of Duracell.
(J) Duncan A. MacInnes (1885–1965) was an American chemist known for his work in electrochemistry and pH detection. He also carried out experimental work to verify Debye-Hückel theory of electrolysis.
(K) Edgar Fahs Smith (1854–1928) was an American scientist who greatly contributed to the development of electrogravimetry (an electroanalytical technique).
(L) Paul Delahay (1921–2012) was an American electrochemist who made outstanding contribution to electroanalytical chemistry.
(M) Søren Peter Lauritz Sørensen (1868–1939) was a Danish biochemist who introduced the pH concept to chemistry and electrochemistry.
(N) William T. Bovie (1882–1958) was an American scientist and inventor credited with conceptualizing the field of biophysics and with inventing a modern medical device known as the Bovie electrosurgical generator.
(O) Jenara Vicenta Arnal Yarza (1902–1960) was the first woman to hold a PhD in chemistry in Spain. She was noted for her work in electrochemistry and her research in formation of fluorine from potassium biflouride by electrolysis.
(P) Boris Vul'fovich (Vladimirovich) Ershler (1908–1978) was a Russian electrochemist studied chemical sources of electrical power, electrical double layer, and theoretical treatment of impedance spectra. Particularly, he was a coauthor of Randles-Ershler admittance model extensively used in modeling of batteries, fuel cells, sensors, etc.
(Q) Paul Dillon (1921–2004) was an American electrochemist—highly important expert in corrosion.
(R) Ferdinand Friedrich Reuss (1778–1852) was a Russian (German origin) chemist who discovered electrophoresis and electroosmosis.
(S) Mikhail Vladimirovich Smirnov (1918–1998) was a Russian/Soviet expert in the electrochemistry of molten and solid electrolytes.
(T) Sergey Vasilievich Karpachev (1906–1987) was a famous Russian scientist, one of the founders of high-temperature electrochemistry school.

INDEX: NAMES

The numbers correspond to the numbers of the mini-sections. The names are organized in the alphabetic order according to the last names of the scientists. Names of people with (*) are only briefly mentioned in the sections devoted to other scientists.

INDEX: THINGS (DEVICES, PROCESSES, CONCEPTS, INVENTIONS, ETC.)

www.ingramcontent.com/pod-product-compliance
Lightning Source LLC
LaVergne TN
LVHW070853210926
848482LV00016B/927
9798893089370